Operational Culture for the Warfighter

Principles and Applications

MARINE CORPS UNIVERSITY
QUANTICO, VIRGINIA
2008

Dr. Barak A. Salmoni
Research Analyst
RAND Corporation

Dr. Paula Holmes-Eber
Professor of Operational Culture
Marine Corps University

Marine Corps University Press
3079 Moreell Avenue
Quantico, Virginia
22134
www. Tecom.usmc.mil/mcu/mcupress/

Fourth Printing, 2009

For sale by the Superintendent of Documents, U.S. Government Printing Office
Internet: bookstore.gpo.gov   Phone: toll free (866) 512-1800;   DC area (202) 512-1800
Fax: (202) 512-2104 Mail: Stop IDCC, Washington, DC 20402-0001

ISBN 978-0-16-083085-3

Contents

Part II
Five Operational Culture Dimensions for Planning and Execution

Part III
Toward Applying Operational Culture

Foreword

Today the U.S. and our allies face a complex but earnest threat. Certainly combat will continue to demand timeless warfighting qualities of initiative, aggressiveness, combined arms skills, ethical decision-making, and more.

Yet, history tells us that a military force unwilling to change, wanting to fight in the old-fashioned way, is doomed to defeat, regardless of the bravery of its soldiers.

Our Marine Corps has proven to be highly adaptive. During Chesty Puller's lifetime, our Corps shifted from what can be described as naval infantry dispersed in ships' detachments, to trench warfare assault troops, to small wars practitioners, to combined-arms amphibious assault troops, to extended land operations alongside our comrades in the Army, to counterinsurgency troops—always maintaining the Corps' ability to defend our country.

This adaptation continued in the 1990s as Marines anticipated and prepared for the three-block wars we have been fighting in Iraq and Afghanistan since 2001.

Fundamental to our adaptation to today's conflict will be the intelligent initiative of all Marines when the enemy hides among innocent people. This demands a keen understanding of culture—the sort of skill practiced by Chesty and his shipmates in the jungles of Haiti and Nicaragua when they served as advisors to non-U.S. forces.

Furthermore, in today's information age, we must recognize that the essential "key terrain" is the will of a host nation's population. This has been demonstrated by our troops in al-Anbar, Iraq, and permits us to gain the trust of skeptical populations, thus frustrating the enemy's efforts and suffocating their ideology.

This superb textbook, a collaboration between our Corps' Center for Advanced Operational Culture Learning and the Marine Corps

University, enables today's Marines to continue their never-ending adaptation to war, keeping our Marines at the top of their game and able to confront and defeat our enemies. Integrating operational cultural principles into the specific conditions where Marines operate in the future will bring depth to touchstones such as "No better friend, no worse enemy," and "First, do no harm." Culturally savvy Marines are a threat to our enemies, so study, challenge, and implement the principles you study in this text. Your buddies, your Nation, and our way of life call on your ability to adapt in the finest traditions of Chesty Puller.

J. N. MATTIS
General, U.S. Marine Corps

Acknowledgments

The authors wish to acknowledge the support and encouragement of several people in the preparation of *Operational Culture for the Warfighter*. These include Gen James N. Mattis, LtGen Keith Stalder, LtGen Richard Zilmer, BGen James Laster, BGen James Toolan, BGen David Garza, and Dr. Jerre Wilson. Extensive discussions with LtGen Joseph Dunford, LtGen Doug Stone, MajGen Mastin Robeson, BGen David Reist, BGen Michael Shupp, BGen W.B. Crowe, Col Matthew Lopez, Col Brennan Byrne, Col Gregg Olsen, Col Paul Kennedy, Col (ret.) Daniel Hahne, Col (ret.) Jerry Durrant, Col (ret.) Steve Fisher, LtCol (ret.) Nicholas Vukovich, LtCol Athony Abati USA, LtCol Jan Horvath USA, LtCol Michael Eisenstadt USAR, Maj Seth Folsom, Maj David Vacchi USA, Maj Remi Hajjar USA, Maj Joseph Lee, Capt Matthew Danner, Capt Barrett Bradstreet, Capt Seth Moulton, Capt Jason Goodale, and MGySgt (ret.) Richard McPherson, led to the germination of ideas that became sections of this book. We also acknowledge the insights of international colleagues, to include Col Bill Monfries and LtCol Daryl Campbell, both of the Australian Army; Dr. Karen Carr, director of the United Kingdom Defence Academy's Centre for Human Systems; and Dr. Charles Kirke, also at the U.K. Defence Academy.

Several individuals read drafts of this book, providing constructive criticism and input. These include BGen Richard Lake, Col (ret.) Jeffery Bearor, Dr. Montgomery McFate, Dr. Kerry Fosher, Col (ret.) Henri Boré French Marines, LtCol Pat Carroll, and Adam Sikes, in addition to members of the senior subject matter staff of the MAGTF Staff Training Program. We wish to acknowledge the editorial assistance of Wendy Overton, Dr. Patrice M. Scanlon, and Stase L. Rodebaugh, as well as the stewardship of this project by the Marine Corps History Division director, Dr. Charles Neimeyer.

Finally, we extend our gratitude to Marine Corps University and the students of its schools for reading and providing essential feedback on an earlier draft of *Operational Culture for the Warfighter*. Their insights and experiences proved fundamental to the book's preparation.

Introduction

Marines know better than anyone else that the challenges of today and tomorrow are more diverse than ever before, running the gamut from high-intensity, combined arms combat to peacekeeping, democracy building, and disaster relief at home and abroad. The diversity of these challenges is accompanied by their simultaneity; Marines have to deal with many different kinds of operating conditions and requirements, all at the same time.

In the 1990s, Marine leaders referred to this diversity of simultaneous actions as a "three-block war."[1] Since then, commanders and strategists have spoken of "irregular warfare" and irregular threats, to emphasize that American armed forces, and Marines in particular, will confront challenges very different from the force-on-force battles of the twentieth century.[2] In the twenty-first century, formal states and regular armies will no longer dominate armed conflict or monopolize coercive force. Our enemies are just as likely to be insurgents whose networks cross national boundaries; warlords who dominate portions of a country or several countries; or international drug and human trafficking cartels. Likewise our potential partners may include foreign militaries or police; local tribal leaders; or people whose long-term ideological agendas differ from ours, but whose near-term interests provide opportunities for pragmatic partnering.

As our experiences in Iraq and Afghanistan have taught us, war in the twenty-first century will require a U.S. military that is as capable of operating through successful enduring relationships with local sheikhs as it is at combined arms operations. Our wars will be "wars *amongst* the people"—not wars *against* the people, and not wars *oblivious* to people.[3] Particularly in the operating environments of today and tomorrow—characterized by insurgencies and political structures under threat from non-state actors—long after hostile military hardware has been destroyed, long after the "high ground" has been taken, Marines will continue to operate with and around people. **The quality of our relationships with**

1

people, in and out of uniform, is of paramount importance in determining mission success. As a master of counterinsurgency (COIN) theory has affirmed, "the battle for the population is a major characteristic of revolutionary war [COIN]."[4]

In order to succeed in the current operating environment, therefore, Marines need to be able to work in foreign Areas of Operation (AOs) with peoples from cultures that are significantly different from our own. Interaction with a growing variety of peoples makes understanding culture a basic component of operational planning, training, and execution. Simply put, the Marine Corps needs Marines to creatively think of, understand, and employ what we call **"operational culture."**

Purpose of Operational Culture for the Warfighter

This textbook is designed to help Marines link concepts of culture to the realities of planning and executing military operations around the world. The book has three primary goals:

- To provide a theoretically sound framework of five basic cultural dimensions, based on clear, academically accurate definitions, which are relevant to military missions.

- To apply these basic cultural principles to actual environments to which Marines and other members of the U.S. military have deployed, or may deploy in the future, showing how the principles of Operational Culture can be applied across the geographic and kinetic spectrum of operations.

- To develop a capacity among Marines at all levels to think systematically about culture, and to apply that thinking to learning about culture in both professional military education and pre-deployment training.

This book is of equal utility to company-grade and field-grade officers, as well as for staff non-commissioned officers (SNCOs). It is to be used in the classrooms of the Marine Corps, and will also serve as a reference tool in the fleet. Finally, it can help commanders and planners to incorporate culture in every step of the road prior to and during deployment, from unit training, to planning, to the ongoing commander's evaluation of the battlespace. These capabilities are particularly germane in current and future operating environments.

Framing the Problem: "Irregular" Warfare and the Significance of Culture

In recent years, "Irregular Warfare" (IW) has been the framework through which the U.S. and allied militaries have thought about military operations. However, a Marine with fleet experience since the end of the Cold War is likely to consider the "irregular" to be the regular mode of operating. This emerges from current definitions of IW, which

> ...has as its objective maintaining or undermining... the credibility and/or legitimacy... of a political authority by the application of indirect approaches and non-conventional means to defeat an enemy by subversion, attrition, or exhaustion rather than [through] direct military confrontation,... though it may employ the full range of military and other capabilities to seek asymmetric advantages, in order to erode an adversary's power, influence, and will.[5]

As a Marine reader will see, these modes and domains of military activity are "irregular" only when compared to traditional doctrines and uses of armed force.[6] Likewise, a Marine with experience in the operating environments of the past decade will understand that **no matter where operations are located on the spectrum of violence, they are about people. Hostile, neutral, or friendly, people are the center of gravity in what militaries do.** As recent joint doctrine attests, IW "is about winning a war of ideas and perception. Its battles are fought amongst the people and its outcomes are determined by perceptions and support of the people."[7]

In this operational mode, Marines of all ranks and billets will face simultaneous, diverse challenges. Among these diverse challenges, the human element will be at least as important as technological and physical elements. In recognition of this, today's senior Marine leadership conceives of the Corps' operational domain as "hybrid wars." Here, Marines will continue to be "the world's finest expeditionary warriors," and will need to combine "equal parts tenacity, courage, and agility" with "the cultural awareness to excel" in the most demanding situations.[8]

To do this effectively, Marines of today and tomorrow need to combine existing know-how and experience-based common sense with a new framework which integrates culture in thinking, planning, and operating at all levels.

Recent history also attests to the centrality of culture in operations. Since WWII, all major U.S. engagements, at all parts of the kinetic spectrum, have involved people in or from regions different from the United States. In Korea, U.S. forces fought against Asians and alongside Asians. In Vietnam, we fought against Southeast Asians, and alongside various peoples from that region. In the operations of the 1980s, in Lebanon, Grenada, and Panama, U.S. military personnel always functioned in areas more or less different from those to which Americans are accustomed. Operations Desert Shield, Desert Storm, and Provide Comfort, as well as deployments to the Balkans and the Caucasus, all signaled what Operations Enduring and Iraqi Freedom drove home: **The U.S. military operates mostly among people, and these people live in environments different from what Americans usually encounter.**

In fact, from the Barbary Pirates campaign, through the Banana Wars, up through current operations in the Horn of Africa, Iraq, and Afghanistan, Marines have always operated in foreign cultures. It is not enough, however, to recognize that culture is a factor in military operations. Rather, in order to succeed in the expeditionary environment of today and tomorrow, Marines at all levels have to understand culture in a military sense.

Back to Clausewitz

To apprehend the centrality of culture to warfare, one can begin by considering Karl von Clausewitz, perhaps *the* towering figure of military thinking in the past two centuries. In the years of the Napoleonic wars of conquest—what we might consider the modern era's first global war against an ideological movement of terrifying proportions for existing states—Clausewitz strove to save Prussia from the French onslaught. In this respect, he may be considered a Foreign Area Officer. Donning the Czarist Army's uniform after the surrender of the Prussian armies, he worked with the Russians, then went on to advise the Prussians while still in Czarist uniform. Marines can also think of Clausewitz as an irregular warfighter: from Konigsberg in 1812-13, he armed the local population as a popular, semi-guerrilla army to fight the French.[9]

Having engaged in warfare nearly continuously from 1792 to 1815, it was as a military thinker reflecting on these years that Clausewitz became most useful to us now. What occupied his thinking was how the French military could fight so successfully and continuously under Napoleon. After all, it was not equipped as well as the Germans, was not generalled as well at levels subordinate to Napoleon, and did not rely on a socially elite, professional force, depending instead on ill-trained conscripts. And, it took on Russia, Prussia, and Britain, among other armies.

Inspired by Gerd von Scharnhorst, his mentor at the Prussian War College, Clausewitz came to the conclusion that the French military success was produced by changes in French society and culture since 1789. The emergence of the ideas of nationalism and citizenship had so motivated French soldiers, that their new identities and motivations had propelled them to great sacrifice and victory at war.

In short, French culture produced French fighting, and in order to understand one's enemies, factors beyond traditional military concern needed to be considered. These included history, beliefs, ideology, demographic changes in social structure, and physical as well as environmental resources: the culture of a people. And, though Clausewitz did not use the word "culture" in particular, the human aspect and attitudes of the population were significant enough for him to include them as a member of his "remarkable trinity" constituting warfare: the people, the army, and the government. Since any warfighter "who ignores any one of them... would conflict with reality to such an extent... [as to] be totally useless,"[10] Marines must study the human environment—culture— and incorporate it in operational planning and execution.

Integrating Culture in the Military Domain

As Marines increasingly engage in multi-national operations and train friendly foreign militaries, **our interaction with a growing variety of foreign peoples and societies will require us to make understanding culture a basic component of training, education, operational planning, and mission execution.** In order to integrate culture as a basic component of Marine activities, we must first clarify thinking and then apply it to the military domain in a way attuned to the operational needs of Marines at all levels. The alternative is to miss out on the "so what" of culture and fail to make it an operationalized concept applied on the battlefield.

The Marine Corps cannot tolerate that kind of failure, because our national security for the foreseeable future requires Marines to understand the cultural dynamics in different corners of the world, and to operate in and among foreign environments with peerless professionalism.

This capability is not, in principle, at odds with existing military doctrine. In preparing for military operations, there are standard considerations during the mission planning phase. These come under the rubric of METT-T: **M**ission, **E**nemy, **T**roops and support available, **T**errain and weather, and **T**ime.[11] In some cases, particularly in urban environments, METT-T becomes METT-TC, for **C**ivilian considerations. This approach has served American warfighters well.

In a world of COIN, Foreign Internal Defense (FID), Disaster Relief, etc—all those things that add up to Irregular and Hybrid Wars where "the population... becomes the objective"— METT-TC becomes most useful if we recast it slightly.[12] Thus, **C** becomes "civilian **cultural** considerations." Additionally, Marines must consider **MC— the "military culture"** of forces against and alongside which they operate.

The Irregular Warrior must therefore plan and operate while accounting for the whole spectrum of METT-TC-MC considerations.

Approaches to doctrine and planning re-emphasized in the past several years enshrine this need. Students in schools, and planners at the regimental through Marine Expeditionary Force (MEF) level, now think in terms of DIME—**D**iplomacy, **I**nformation, **M**ilitary, and **E**conomics—or, particularly at the MEF and combined, joint levels, PMESII—**P**olitical, **M**ilitary, **E**conomic, **S**ocial, **I**nfrastructure, and **I**nformation. No mater what the "alphabet soup," all these elements are embedded in local cultures. As Lines of Operation (LOOs), it is impossible to pursue them in an integrated, holistic fashion absent an understanding of cultural dynamics.[13]

Just as doctrine has a place for culture, the term "Operational Culture" itself is not entirely new. Though the term has not been prominent in the social sciences, the well-known sociologist Ward Goodenough used it to refer to a set of norms and behaviors that people switch into, or activate, given the group they are in for any given purpose. People thus "operationalize" whatever culture they need.[14] This already hints at an important factor in the military domain: **Culture for military people is useful only if we can render it operationally relevant.**

This text will define "operational culture" in subsequent chapters, presenting a flexible way of thinking about a continually changing operational environment. As a start, however, Marines need to understand that **Operational Culture as a skills set comprises:**

- **Knowledge of considerations at play in human societies globally.**

- **Capability to function among diverse peoples.**

- **Understanding of specific human societies to which one deploys.**

- **Ability to successfully integrate this knowledge and understanding into operational planning and mission execution.**

Purpose and Structure

This text is intended to orient Marines' minds to operational culture in the context of expeditionary and irregular operations: the bread and butter of what Marines do.

It seeks to do so by furnishing a framework for thinking about how "people-as-center-of-gravity" can be a factor in military operations. We have chosen the following structure:

Part I: Terms of Reference

Chapter 1 puts "operational culture" in the context of historical, current and emerging military challenges, and examines the utility to the Marine of cultural anthropological approaches.

Chapter 2 questions some commonly accepted definitions of culture, suggesting the difficulties they present to military users. It also provides a more precise, conceptually sound set of definitions related to culture, appropriate to the military profession.

Part II: Five Cultural Dimensions for Marine Operations

Chapter 3 starts the process of explaining the Five Operational Culture Dimensions for planning and operating, with a focus on the physical environment.

Chapter 4 explores the economic dimension.

Chapter 5 looks at social structures applicable to the warfighter.

Chapter 6 examines political structures.

Chapter 7 considers the impact of beliefs and symbols on operations.

Part III: Towards Applying Operational Culture

Chapter 8 looks at the linkages among these five dimensions, through examination of case studies.

Chapter 9 develops a framework for culture learning in education and training, according to accepted taxonomies of learning.

The Conclusion ties definitions and dimensions of culture together by looking at militaries as cultures, and by showing the congruence of Marine Corps doctrinal principles with those of Operational Culture.

As We Cross the Line of Departure...

As we begin this exploration of cultural principles, dimensions, and applications in the military domain, we need to bear in mind two major considerations. First, understanding foreign cultures is by nature difficult. Becoming competent as an actor in a foreign environment is even more challenging, requiring much study and time. It is not a perfect science, and there are no guarantees about what understanding some aspects of a culture portends in terms of success—military or otherwise—in that culture. Studying concepts, tools, and considerations as we do in this text, however, remains a core part of preparation for the irregular warfare world. This is because in future military operations, Marines will need to orient their minds to the human aspects of the battlefield, to make cultural knowledge truly "operational."

Second, this text is a learning tool, for use in schools of the Marine Corps and as a reference for reflection in the fleet. It is not a doctrinal publication directing you how to act or prescribing procedures. As such, it is not a quick or simple read. The broad and deep phenomenon of how diverse human realities influence—and are influenced by—military operations cannot be reduced to a quick and easy read. Consider the words of Clausewitz in this regard, who felt we must avoid the temptation of looking for a "kind of truth machine" allowing Marines the "mechanical application of a theory."[15]   **This book is designed as an educational text: Read slowly, think much, and derive conclusions that you can implement in an operational environment characterized by diversity and uncertainty.**

Notes

[1] Gen Charles C. Krulak, "The Strategic Corporal: Leadership in the Three Block War," *Marines Magazine*, January 1999.

[2] U.S. Marine Corps Combat Development Command and U.S. Special Operations Command Center for Knowledge and Futures, *Multi-Service Concept for Irregular Warfare* (version 2.0, August 2006), at http://www.mcwl.usmc.mil/. For the doctrinally approved definition of IW, see *Irregular Warfare Joint Operating Concept* (IW JOC) Version 1.0, 11 September 2007: IW is "a violent struggle among state and non-state actors for legitimacy and influence over the relevant populations. IW favors the indirect and asymmetric approaches, though it may employ the full range of military and other capabilities, in order to erode an adversary's power, influence, and will."

[3] See Gen Rupert Smith, *The Utility of Force: The Art of War in the Modern Word* (New York: Knopf, 2007), Part III in particular.

[4] David Galula, *Counterinsurgency Warfare: Theory and Practice* (Westport, Conn: Praeger Security International, 1964 [2006]), 4.

[5] See *Multi-Service Concept for Irregular Warfare*, 5-6.

[6] For the regularity of "irregular" warfare, see Barak Salmoni, "The Fallacy of 'Irregular' Warfare," *Royal United Services Institute Journal*, August 2007.

[7] *Multi-Service Concept for Irregular Warfare*, 5-6.

[8] LtGen James N. Mattis (USMC) and LtCol Frank G. Hoffman (USMCR, ret), "Future Warfare: The Rise of Hybrid Wars," *U.S. Naval Institute Proceedings*, November 2005, 18-19.

[9] Peter Paret, *Clausewitz and the State: The Man, His Theories, and His Times* (Princeton, NJ: Princeton University Press, 2007).

[10] Karl von Clausewitz, *On War*, ed. and trans. by Michael Howard and Peter Paret (Princeton, NJ: Princeton University Press, 1976), 89.

[11] See MCRP 5-1a: *Doctrinal References for Expeditionary Maneuver Warfare* (HQMC, 2002). 59-66 in particular.

[12] David Galula, *Counterinsurgency Warfare*, 74.

[13] See Joint Pub 5-00.1: *Doctrine for Campaign Planning* (25 January 2002); Joint Pub 5-00.2: *Joint Task Force Planning Guidance and Procedures* (13 January 1999); Joint Pub 5-0: *Joint Operation Planning* (26 December 2006); Robert D. Steele, *Information Operations: Putting the "I" Back into DIME* (Carlisle Barracks, PA: Strategic Studies Inst. of the Army War College, 2006); Kris A. Arnold, *PMESII and The Non-State Actor: Questioning the Relevance* (Ft Leavenworth, KS: Army Command and General Staff College – School of Advanced Military Studies, May 2006); Maj Robert Umstead and LtCol David R. Denhard, USAF, "Viewing the Center of Gravity through the Prism of Effects-Based Operations," *Military Review* Sept-Oct 2006, 90-95.

[14] Ward Goodenough, *Culture, Language, and Society* (Reading, MA: Addison-Wesley, 1971).

[15] Karl von Clausewitz, *On War*, 168, 150.

Part I

Terms of Reference

Upon visiting or deploying to a foreign country in a military capacity, it might be tempting to fall victim to two different ways of initially coming to terms with the unfamiliar area. One method is to *mirror-image*: to assume that things do—or worse yet, should—operate just as they do "back home," and that the people do—or again, worse yet, should—act like Americans do, and interpret events just as would Americans.[1] Mirror-imaging, in effect, involves blindness to difference from culture group to culture group.

The second temptation is to view understanding the new culture group as an impossible project. In this approach, "difference from" comes across as the most prominent, defining characteristic of foreign areas: the people speak a different language; they have unfamiliar customs; and they seem to act in unpredictable ways. Symptoms of this view include avoidance of cultural dynamics as a consideration—"I can't do anything about it, so why care;" overfixation on culture as a hot potato—"They're so different and I still can't do anything about it, so I have to watch out;" or simply limited faith in the possibilities of inter-cultural bridging—"I'm me, he's him, and never the twain shall meet."

In this book, we offer a third, more productive approach to culture: viewing culture as a logical system which can be understood using theories and principles developed during more than 150 years of research and study by social scientists. Although people are, by nature, variable and unpredictable, they still need to work with others in social and cultural groups. These groups—and their associated beliefs and structures—are organized according to logical, understandable principles that every person living in the culture must understand, at least intuitively, in order to get along with each other. With some basic study, Marines can also recognize and understand these principles and apply that understanding to their operations.

This book does not assert that culture is linear or predictable or that militaries should expect culture learning to provide instant answers or formulas. But, in the following chapters, we do offer Marines fundamental tools and conceptual models to understand, operate in, and when possible, influence his or her area of operations. Part I of *Operational Culture* begins by presenting basic terms of reference for understanding culture.

In Chapter One, we discuss the influence of culture in historical military operations, and also examine attempts throughout history to understand culture in a way useful to each era's expeditionary operations. In Chapter Two, we move onto a discussion of useful and unprofitable ways to think about culture and culture groups from the standpoint of military operations. This prepares us to define fundamental terms and core terms of reference at the end of Chapter Two.

Note

[1] For more on mirror-imaging, see Herbert Kelman, "Social-Psychological Dimensions," in *Peacemaking in International Conflict: Methods and Techniques* edited by I. William Zartman and J. Lewis Ramussen (Washington, DC.: United States Institute of Peace, 1997); Daniel Druckman, "Social-Psychological Factors in Regional Politics," in Werner Feld and Gavin Boyd, eds., *Comparative Regional Systems: West and East Europe, North America, The Middle East and Developing Countries* (New York: Pergamon Press 1980); Frank Watanabe, "How To Succeed in the DI: Fifteen Axioms for Intelligence Analysts," *Studies in Intelligence* (1997): http://www.cia.gov/csi/studies/97unclass/axioms.html. Also see "Peacebuilding: Social Psychology," part of the School for International and Area Studies Conflict Management Toolkit, located at: http://cmtoolkit.sais-jhu.edu/index.php?name=pb-socialpsych.

Chapter 1

Context

The Introduction to *Operational Culture* demonstrated the centrality of culture to current and future operations, and indicated specific skill sets required for Marines to integrate culture into planning and operations. This may suggest that "culture" is a new area of thinking and study for the military. In light of the last two-hundred years of military history, however, that is not the case. Rather, as we will show in this chapter, military operations of the past two centuries have been a driver of interest in foreign cultures, both in academia and military circles. In fact, one could argue that there had been a military-cultural studies partnership that endured from the nineteenth century until the 1960s. This partnership produced advances in the field of anthropology, but also led to certain simplistic models and misunderstandings of actual cultural dynamics in foreign areas.

After surveying these developments, we will introduce a more rigorous conceptual approach to cultures which can serve Marines' needs today and in the future. Based upon accepted anthropological models but adapted to concerns of operators, this approach amounts to a framework for examining cultures in a military context. This framework is called the Five Operational Culture Dimensions, which permit the Marine to understand global cultures in a fashion suited to mission planning and execution.

Before we can begin this discussion, however, we need to establish a basic definition of "operational culture." This initial definition will prepare us to develop a more comprehensive one in subsequent chapters.

Operational Culture: Those aspects of culture that influence the outcome of a military operation; conversely, the military actions that influence the culture of an area of operations.

Though in this book we focus primarily on the first part of the definition—"aspects of culture that influence military operations"—both aspects of operational culture are important, and we will incorporate them both into our discussion when appropriate.

A Historical Approach to Culture in Operations

As Marine leaders speak today of IW, Distributed Operations (DO),[1] and the importance of culture to warfighting, **we must remember that it is not the first time militaries have done something new—or considered human and cultural factors in planning and operating.** In fact, at various times in history, armies have engaged in what, compared to previous generations, appeared "irregular." The first time Greek armies faced a phalanx, that was irregular compared to past experiences; the operations of Caesar's or Augustus' armies in Gaul and the Rhineland were, compared to Rome's past, "distributed" operations. Every generation has a way of warfare, and every era has a revolution in military affairs.[2] Likewise, as we saw above with Clausewitz, these changing ways of thinking about and executing operations include considering human and cultural elements.

Especially over the past 250 years, as European militaries expanded their operations into non-Western regions such as Africa, Asia, and the Middle East, it became evident that traditional military techniques were not adequate to deal with the new military environment. As a result, British, French, and Dutch operators and thinkers began to include local cultural behaviors and beliefs in their military planning and operations. As military personnel began to operate overseas in a sustained fashion, theories about society began to emerge, linked to problems in overseas Areas of Operation (AOs).

Through the middle of the nineteenth century and continuing into the twentieth century, European scholars followed their armies, or sometimes went with them, collecting data about the peoples among whom their armies operated. Often ancient history was attractive, as it could supposedly "capture the soul" of a foreign people. This is one reason that the French Institute d'Egypt emerged in the wake of the French Expedition to that country under Napoleon.

By the beginning of the twentieth century, then, it was accepted that to know and command an AO, one had to understand the people in that area. The more that colonial officers and overseas administrators got into foreign environments, the more they found that they were outnumbered compared to indigenous warriors and peoples, and had to work "in the system," and make the local people function for them.

This was the era when, in North Africa, the Levant, South Asia, Central Africa, and Indochina, British and French officers at the tip of the colonial spear confronted disadvantageous troop-to-task ratios. In only very few places did European military leaders attempt to replace local leaders. Instead, they relied on local points of contact.

In order to do this, they needed to know very basic, structural information: Who in this town is the headman? What does the tribe look like? How does justice work here? Who are the medicine men or ritual leaders? In short, **military personnel** and civilian administrators charged with providing stability and security in foreign areas **needed to know how**, in order to meet economic, political, military, and psychic needs, **cultures organized themselves into institutions:** What were the structures, how did they function, and how did colonial officers imagine they could "guide" and "manipulate" the cultural environment?

As the uniformed and civilian irregular warriors of the early twentieth century worked out the nuts and bolts of these matters on the ground in far off lands, European scholars, often acting in conjunction with colonial powers, turned attention to the kin and social structures of local cultural groups as well as trying to understand their symbols, religions and rituals.[3] Anthropological studies of communities—termed ethnographies—began to emerge on these same topics, implicitly inspired by the challenges their soldiers faced. This is when field work yielded detailed studies of groups, leadership practices, power structures, and economic networks among foreign peoples—almost always assumed to reside in "far-off," "exotic" lands.

In the meantime, the U.S. military was facing its own cultural challenges, as it grappled with the problems of settling the west and the resulting wars with Native American groups. U.S. scholars, faced with a confusing myriad of Native American languages and cultures, began to study the symbolic meanings of words in languages.[4]

Later, particularly from the 1920s through the 1960s, when the era of insurgency and COIN was in full swing, the ideologies of ethnicity, nationalism, and third-worldism motivated the largely successful efforts to break free of colonial rule. In fact, just as the nineteenth century had been the period when Europe had witnessed the era of nationalism, the post-WWI era saw nationalism explode in the non-European world.

Policy makers in a world of burgeoning nation states then began to ask how different cultures might affect how people in different nations acted. Was there such a concept as a "national culture" or "national character?" If so, what did that mean for issues of diplomacy, politics, economic policies, approaches to warfare, and even tactical preferences?

Working with the military in WWII, anthropologists embarked upon a sadly misguided program to simplify cultures into an easily predictable system. This approach, later termed the study of "Culture and Personality," attempted to reduce the complexity of culture by proposing that cultures create people with similar mindsets or personalities. The effort to profile "national personalities" or "national characters" resulted in simplistic and stereotypical explanations for the way people acted in other nations.[5] Such studies produced overly generalized descriptions of cultural groups resulting in sweeping statements such as:

- Japanese children's swaddling and toilet training led to a national Japanese character of internal repression of personal desires and subservience to elders.[6]

• Due to the nature of the Arabic language, Arabic culture is given to formulaic, rhetorical expression, so that as thinkers and actors Arabs are less capable of concreteness and commitment-fulfillment.[7]

• Russians are intransigent negotiators because Russian swaddling (!) techniques make them tight-fisted.[8]

Although "Culture and Personality" studies were immensely popular in the 1950s, anthropologists soon realized that blanket statements about a cultural mindset failed to capture the important social structures, belief systems, and competing agendas of people within a cultural group. The "Culture and Personality" approach was then abandoned for more sophisticated research that focused on the complex and evolving nature of cultures, providing a richer and more accurate view of the way that people in diverse cultures viewed the world. This new approach coincided with a parting of the ways between anthropologists—who became more concerned with understanding cultures from a policy and development point of view—and the military, which focused its energies on a high tech "star wars" approach to warfare as the result of the Cold War.

Unfortunately, in the general population, as well as in the U.S. military, the idea that "culture = mindset = national psychology" continued to be popular long after anthropologists had turned to other more fruitful studies. The military's continued interest in the idea of a cultural mindset reflected its belief that somehow culture itself and the people within a culture were a code to be cracked. Military efforts to understand foreign cultures continued to work on the premise that if one cracks the code, one can then work the culture like a computer program to determine peoples' responses to our actions. Just like there is a Microsoft operating system different from the Macintosh operating system, the "culture = mindset = national psychology" framework reinforced the fantasy that there is an "Arab operating system," an "Asian operating system," or a "Western operating system."

One very popular psychologist who has applied the "Culture and Personality" approach to business and management studies is Geert

Hofstede. His cross-cultural studies on different management styles and their relationship to national cultures began in the 1960s and have continued up to recent years.[9] Using survey responses of IBM business leaders, he constructed a model of management styles as they relate to a simplified schema of "national culture." His work posits a list of psychological dimensions that differentiates business leaders around the world: individualism ↔ collectivism; power distance; uncertainty avoidance; relationship focus ↔ deal focus; and long term short ↔ term time orientation.

Hofstede's studies are grounded in psychology, not anthropological or sociological studies of foreign cultures. His work compares the psychological predispositions of business leaders, and the differences in their management styles. This is all a *far cry from explaining an entire culture and the people in it.* Furthermore, his dimensions were posited to explain psychological difference *along a continuum*; not offer black and white dichotomies. For example, cultures are not *either* individualistic *or* collective, but lay somewhere along a continuum between these two extremes: hence our choice of "↔" and not "vs." in the previous paragraph.

However, because a "this or that" approach fits easily with the U.S. military's "go or no-go" schema, military people have read Hofstede's analysis in black and white, "yes-or-no" terms. As a result, Hofstede's dimensions have become almost as a prescriptive checklist, and a misinterpretation of it has spread across the services. His and other similar approaches have therefore gained a deceptive appeal as a "silver bullet" to understanding foreign peoples.

Unfortunately, Hofstede's dimensions—which focus on specific personality traits in business—offer only limited applicability to a Marine who is working with local people in a foreign AO, whose experiences and concerns are radically different from the business people in Hofstede's studies. Hofstede's dimensions focus on human traits, themes, and attitudes, but do not examine or account for the main drivers of social, economic, and political life and belief systems for culture groups. In effect, concepts such as collectivism vs. individualism, or uncertainty avoidance, may have relevance in certain situations such as understanding cross-cultural negotiations or child-raising practices, but they do not describe the fundamental challenges that drive the lives of people in foreign cultures.

In fact, misuse of Hofstede and similar psychological studies, when rendered simplistically, produces prejudices and cultural stereotypes rather than better cultural understanding. Such stereotypes disadvantage Marines in a foreign environment, making it difficult to respond to the actual situation and individuals at hand. This is because "this vs. that" mindsets tempt people to react to new situations with a formulaic understanding obscuring important details. Put simply, fixation on "the Arab mind" will obscure the character of the real Arab with whom one deals. Likewise, focus on the "Latino mentality" makes it difficult to get to know the real Latin American in front of a Marine.

All this will undermine mission success. This is because while in the information age culture seems to be the "collective programming of the mind" and the "mental software" of a "national culture,"[10] **Marines must not consider people programmed into cultures, and thus locked into certain kinds of behaviors. Such an approach creates simplistic stereotypes** based on misplaced ideas of a "typical American" vs. "median Arab" or "average Middle Eastern male."

This is a problem because it could cause Marines and others operating in a region to fail to grasp the true complexity of the cultural situation, resulting in poor analysis and ineffective operational planning. Indeed, in operating environments where culture has been a central factor, such as Bosnia, Somalia, and Iraq, the **messiness of operations has shown this mechanistic perspective to be of little utility to operators,** who are likely to agree that cultures themselves are "messy," or a "fuzzy set," defying patterns.[11]

All of the historical approaches to culture reviewed above have their problems, by either focusing on too much of one thing, or by taking questions to a point of blind, racist belief. Still, thinking about them is useful to the irregular warrior of today, permitting us to sum up what we have encountered thus far: Irregularity in military operations is not new, and thinking about, examining, planning, and operating according to the changing human environment is as much a part of the military profession as is learning to fire a rifle. Further more, when Marines seek to incorporate operational culture into warfighting, they need to do so within an intellectually sound framework.

o

The Conceptual Context

In arriving at a framework to understand cultures globally, the Marine need not arrive at the fight alone. Today, more sophisticated social scientific approaches to culture furnish the Marine with up-to-date, relevant and useful ways of analyzing culture during both planning and execution. From the point of view of the Marine operating in a foreign context, three major anthropological models should be integrated into the irregular warfighter's practice:

- An "ecological model" which focuses on the relationship between cultures and the physical environment.

- A "social structure model" which examines the way that the social structure of a group affects the roles, status and power of the various members.

- A "symbolic model" which studies the beliefs, symbols and rituals of a cultural group.[12]

By integrating these models into warfighting, **the operator can move beyond mere description in order to explain, and if possible influence, human behavior.**

The Ecological Model. This analytical method focuses on the ways that the physical environment affects the economic and social relationships within a cultural group. Emerging from American and British studies of the linkages between the political structure of a society and its economic system, The Ecological Model analyzes the relationship between available physical and social resources (such as water and land, fishing and hunting of game, or even important hi-tech commodities such as oil and the Internet) and the consequent social and political structures of a culture.

Thus, The Ecological Model understands conflict as a logical outcome of competition over important limited resources. Hence, poverty, starvation and underdevelopment are seen as precipitating factors in many of the conflicts in Africa, Asia and South America in the past century. Likewise, land pressures due to an increasing population are linked to many of the wars in Europe in earlier centuries.

The Social Structure Model. This examines how people organize political, economic, and social relationships, and how that method of organization determines roles, rights, and privileges of a group's members. Most societies do not distribute power and resources evenly. The Social Structure Model thus allows the culture operator to evaluate which groups hold authority and power in an AO.

In particular, The Social Structure Model can pinpoint:

- Tenuous alliances that generate power structures

- The rivalries within these alliances

- Changing dynamics of power within and among groups

- Challenges to existing power structures, either by included or excluded groups within a society

As such, unlike The Ecological Model, The Social Structure Model explains conflict and war, as Marines see it from the theater- to the company AO-level, in terms of the structure of a society. In this model, losing groups (whether an ethnic minority or social class) battle existing systems in order to negotiate better access to goods or to shift their position within the power structure. The Social Structure Model is particularly instructive in understanding insurgencies, such as the current situation in Iraq, in addition to most of the contingencies into which Marines have been drawn over the last two decades.

The Symbolic Model. Beginning in France in the middle of the twentieth century, in recent years this approach has become popular in the U.S. as well. For The Symbolic Model, physical geography and social structures do not illuminate central features of a culture. This paradigm, or method, concentrates on culture's symbolic or ideological nature. In this model, culture is the product of thought and humanly constructed belief systems and values. Human behavior is therefore explained through ideals which guide a person's choices.

Due to its ideological nature, The Symbolic Model has been particularly suited to the analysis of religion, as well as to the study of semantics. It can help to explain the symbolic, psycho-emotional meaning of concrete, practical reasons behind conflict. In short, this model can explain why problems that could be solved "if only people would sit down and talk" do not go away. It shows how factors such as history, identity, and the symbolic meaning of space and events can become part of—if not the fixation of—negotiations, replacing dialogue.

In the realm of military engagement, consider how each of these three pillars of the thinking and planning framework for operational culture would explain conflict:

Models and the Explanation of Conflict		
1	Ecological	War is related to battles over limited resources.
2	Social Structure	Conflict results from vying plays for power in an unequal social structure.
3	Symbolic	War is an inevitable conflict over identity and ideology between competing systems.

The Five Operational Culture Dimensions

Models on their own remain too abstract and general to help the warfighter. These three, however, have the particular military merit of furnishing **five specific cultural dimensions of the battle-space that Marines can examine and incorporate into plan-**

ning and execution. Although we will cover the specific human dimensions of the battlespace relevant to Marines in more concrete detail in Part II, we summarize them below, illustrating their connection to the three models discussed in this chapter.

From Anthropological Models to Operational Culture Dimensions	
Ecological Model	
Dimension 1: **The Physical Environment**	The way that a cultural group determines the use of the physical environment. Who has access to important physical resources (water, land, food, building materials etc.) and how the culture views these resources (e.g. land is owned or free to everyone)
Dimension 2: **The Economy**	The way that people in a culture obtain, produce and distribute physical and symbolic goods (whether food, clothing, cars or cowrie shells).
Social Structure Model	
Dimension 3: **The Social Structure**	How people organize their political, economic, and social relationships, and the way this organization influences the distribution of positions, roles, status, and power within culture groups.
Dimension 4: **The Political Structure**	The political structures of a culture group and the unique forms of leadership within such structures (bands, acephalous societies, councils, hereditary chiefdoms and tribal structures, electoral political systems etc.). The distinction between formal, ideal political structures versus actual power structures.
Symbolic Model	
Dimension 5: **Beliefs and Symbols**	The cultural beliefs that influence a person's world view; and the rituals, symbols and practices associated with a particular belief system. These include also the role of local belief systems and religions in controlling and affecting behavior.

These dimensions will be referred to henceforth as the **Five Operational Culture Dimensions.** These Dimensions are distinctly different from anthropological models recently used by the military, such as Hofstede's construction of dimensions of national culture, which we have discussed above.[13]

Conclusion

This chapter has briefly introduced the Five Operational Culture Dimensions. We will return to them in much more detail throughout this book, as they will anchor our exploration of the relationship between culture and military operations. Before we can do that, however, we will need to look at some existing definitions of "culture" in the next chapter. After critiquing them in terms of their validity for both real human environments and military operations, we will present a series of definitions of culture and associated terms that accord with these five dimensions.

Notes

[1] See "A Concept for Distributed Operations," Dept of Navy, Headquarters USMC, 25 April 2005; Marine Corps Combat Development Command, "Questions and Answers about Distributed Operations:"
http://www.mcwl.usmc.mil/SV/DO%20FAQs%2016%20Mar%2005.pdf; "Marine Air-Ground Task Force Distributed Operations," *Marine Corps Gazette* Oct 2004, 34-36.

[2] Recent expressions of the evolution of forms of warfare can be found in T.X. Hammes, *The Sling and the Stone: War in the 21st Century* (Osceola, WI: Zenith Press, 2004); idem., "Fourth Generation Warfare Evolves, Fifth Emerges," *Military Review*, May-June 2007; Gen Rupert Smith, *The Utility of Force: The Art of War in the Modern World* (New York: Knopf, 2007).

[3] For further reading, see E.E. Evans-Pritchard, *Kinship and Marriage Among the Nuer* (Oxford: Clarendon, 1951); idem., *The Nuer* (Oxford: Clarendon, 1940); C. Levi-Strauss, *The Elementary Structures of Kinship* (Boston: Beacon, 1969); B. Malinowski, *Argonauts of the Western Pacific: An Account of Native Enterprise and Adventure in the Archipelagoes of Malanesian New Guinea* (London: Routledge, 1922); idem., *Magic, Science and Religion* (New York: Doubleday, 1954); A.R. Radcliffe-Brown, *Structure and Function in Primitive Society* (London: Cohen and West, 1952).

[4] For Further Reading, see F. Boas, *Race, Language and Culture* (New York:

Macmillan, 1940); Benjamin Lee Whorf, *Language, Thought, and Reality* (Cambridge, MA: MIT Press, 1956); Edward Sapir, *Culture, Language, and Personality: Selected Essays* (Berkeley: University of California Press, 1986).

[5] For an early presentation of ideas later used in "culture and personality" studies, see Ruth Benedict, *Patterns of Culture* (Boston: Houghton Mifflin, 1934); associated with this perspective are Victor Barnouw, *Culture and Personality* (Homewood, Ill.: Dorsey Press, 1963); and Geert Hofstede, "National Cultures Revisited," *Behavior Sciences Research* 18 (1983), 285-305.

[6] Ibid. (Barnouw).

[7] See Raphael Patai, *The Arab Mind* (Long Island, NY: Hatherleigh Press, 2002 [1973]).

[8] For discussion and critique of these cultural explanations of Russian negotiating techniques, see Jerrold R. Schechter *Russian Negotiating Behavior: Continuity and Transition* (Washington, DC: U.S. Institute of Peace Press, 1998).

[9] See Geert Hofstede's *Cultures and Organizations: Software of the Mind* (Cambridge, UK: McGraw Hill, 1991); and his *Culture's Consequences: International Differences in Work-Related Values* (Sage Publications, 1980).

[10] Ibid.

[11] H. Spencer-Oatey, *Culturally Speaking: Managing Rapport through Talk Across Cultures* (London: Continuum, 2000), 4.

[12] These methods correspond to well accepted analytical approaches in academic anthropology:

"The Ecological Model"	=	Ecological Anthropology
"The Social Structure Model"	=	Structure-Functionalism British Social Anthropology
"The Symbolic Model"	=	The Symbolic Approach

For a more complete discussion of these approaches in academic anthropology, see Alan Barnard, *History and Theory in Anthropology* (Cambridge, UK: Cambridge University Press, 2000); Jerry D. Moore, *Visions of Culture: An Introduction to Anthropological Theories and Theorists*, 2nd ed. (Lanham, MD: Alta Mira Press, 2004); Paul A. Erickson and Liam D. Murphy, *A History of Anthropological Theory, 2nd* ed. (Broadview Press, 2003); Philip Carl Salzman, *Understanding Culture: An Introduction to Anthropological Theory* (Long Grove, IL: Waveland Press, 2001); John Monaghan and Peter Just, *Social and Cultural Anthropology: A Very Short Introduction* (Oxford, UK: Oxford University Press, 2000).

[13] See Geert Hofstede, "National Cultures Revisited;" Ernst Cassirer, *An Essay on Man* (New Haven, CT: Yale University Press, 1944), 65-70 in particular.

Chapter 2

Defining Culture

In the previous chapters we have discussed the importance of understanding culture in military operations; the historical relationship between the military and the study of culture; and theoretical models and a framework of culture that are relevant to military operations. In this chapter we turn our discussion to the actual subject of this book: What is and what is not "culture"? First, we examine several of the ways that military writers have already defined culture, and we evaluate common misconceptions associated with these concepts. Next, we explore some of the ways that anthropologists define culture, developing an operational definition of culture that is both conceptually accurate and applicable to military needs. We focus on several essential characteristics of culture that Marines need to grasp in order to successfully plan and undertake operations in foreign areas of operation. In the final section, we provide definitions of cultural concepts relevant to Marine operations.

Military Definitions of Culture

Currently there are many definitions of culture in use by US military organizations. Most of these definitions reflect understandable preferences for a) simple, generalized abstractions that can be applied quickly to any situation; b) information reducible to easily followed check-lists; and c) presentation of information in ways that conform to ways of knowing and communicating in other domains of military affairs. Definitions of culture meeting these criteria may appear appropriate to the Marine. However, when applied in the field they are not likely to be effective, because they are fundamentally flawed conceptually—in short, they are misleading as to the nature of culture, and its influences on military operations. In essence, these approaches to culture are much like handing a

Marine armored platform that looks great, but has no engine, no brakes, and no steering mechanism.

Below we discuss some of the more common military under-standings of culture and examine their limitations. In the following sections, we then offer more accurate—although perhaps less "comfortable"—definitions of culture.

Many definitions of culture borrowed by recent military writing are so broad as to present abstractions tiptoeing in the direction of stereotypes. Even more important, upon closer examination, many of these definitions are conceptually muddled: mixing terms and concepts without understanding their relationships or meanings. Consider these examples:

- One recent article defined culture as the "origins, values, roles, and material items associated with a particular group."[1] However, calling culture as "origins… of a group" confuses cause and effect, while "values" is too abstract a term. Such def-initions are too theoretical for application to the military con-text.

- Another speaks of culture as the "total shared, learned be-havior of a society or a subgroup."[2] Here, "culture" emerges as so expansive that it is nearly impossible to analyze, consider, plan for, and operationalize.

- In yet another place, we find culture as "an integrated sys-tem of learned behavior patterns that are characteristic of the members of any given society… the total way of life,"[3] and the "outward expression of a unifying and consistent vision brought by a particular community."[4] These constructions imply "cul-ture" is the "what you see" of something that *all* people in a place and time *agree* on, act in terms of, and pass on as a *co-herent whole* to the next generation. The role of individual agency and action is neglected in such a homogenizing ap-proach.

These definitions have the merit of using terms appealing to military readers. Yet, they tend to focus on abstract, conceptual aspects of culture which are difficult for a Marine to observe and evaluate. In terms of current anthropological understandings of culture, such definitions have several serious problems:

1. These definitions tend to present culture as somehow *separate* from people: as if culture is a thing that "exists somewhere out there" forcing people to behave in certain ways. However, culture is the product of human interaction and thought, not a mysterious force with its own purposes and agendas.

2. Underlying these definitions is an assumption that culture is homogenous and static. The definitions imply that there is something called an Arab or Japanese culture which has made all Arabs and Japanese behave the same way for the past 2000 years. This assumption underlies the military mentality that we can somehow create computer simulations which will predict how an Arab or a Japanese person will behave or act. As we discuss below however, culture is far from static and not all people in the same culture act or think the same.

3. The reader of such definitions gets the sense that culture is a magic "black box" composed of strange, abstract terms such as values and ideals or ways of life. Unable to open the black box, the Marine is likely to dismiss culture as indecipherable and hence irrelevant to operations. Although culture is indeed complex, it is not unintelligible. From the point of view of anthropological study, the social relationships, structures, meanings and processes underlying the cultures we observe can be analyzed and explained just as well as the functioning of the components of an engine can be described and predicted.

4. Finally, in the military context, these definitions leave the reader with a "so what, what do I *do* with all this" reaction, because they fail to describe real people, real social dynamics, and the real environment that Marines must navigate.

While the definitions above suffer from too much abstraction, at the other extreme are military definitions of culture which can describe the rocks, the pieces of sand, and the leaves on the plant, but still miss the full picture of the landscape. They provide concrete and easily understandable terms for the military, but provide little conceptual guidance, so that culture simply becomes a huge list of disorganized information of little use to the operator. For example:

- A recently published text meant for military readers defines culture as "a shared set of traditions, belief systems, and behaviors. Culture is shaped by many factors, including history, religion, ethnic identity, language, and nationality."[5] This definition, which is typical of military definitions, confuses broad categories with their components. Like much of the literature being rapidly produced by military writers, culture is defined in terms of a random "shopping bag" list of items that seems relevant to operations, but that has no logical theoretical or conceptual relationship between items.

- A recent military definition of culture characterized it as one among several political-military factors:"…Learned and shared attitudes, values, and ways of behaving in a society. Culture includes customs, folkways, manners, mannerisms, etiquette, behaviors, body language, gestures, celebrations, milestones, dress, outlooks, perceptions, and thought patterns. It is embodied in history, art, myths, legends and heroes. It addresses appropriate responses to situations. It determines the circumstances and quality of apology, retribution, reward, punishment, equity, commiseration, disdain, shame, guilt, congratulations and pride. It selects and applies social sanction and reward. It expresses itself in superstitions, outlooks, perspectives, conventional knowledge and points of view. It encompasses the sense of time, individuality, possessions, sharing, self-worth and group-worth. It establishes the social hierarchy, defining roles by sex, age, position, religion, wealth, family and profession. In essence, culture defines what is and is not okay, accepted, and normal."[6]

Given this kind of a laundry list, planners will struggle, often in vain, to collect accurate data to fill in all the blanks. Ultimately, no matter how detailed the data, a checklist of information on a culture offers very little *conceptual* understanding to a warfighter.

Such a list is not the same as understanding the *meaning* of the data for the people who live in the region.

It is important that Marines distinguish between long lists of cultural data (which may overwhelm the planner with too much irrelevant information) and cultural understanding (which requires the ability to interpret and use the data appropriately). Collecting a long list of cultural facts about history, manners, etiquette, dress and superstitions may look impressive. However, as many Marines today can attest, knowing whether or not one should show the soles of one's feet in Iraq does not provide much guidance in planning operations in a foreign culture. In order to move from "cultural sensitivity" to operational culture, Marines need to **understand the underlying meaning and organization of social relationships and behaviors within a cultural group.**

Information and data might diminish operational culture ability	≠	**Understanding and know-how** can increase cultural situational awareness

In addition to overly abstract or overly concrete definitions of culture, a third way that culture is defined in U.S. military circles is in terms of a very popular military metaphor: the "Human Terrain."[7] "Human terrain" does possess some useful applications at the tactical through strategic level. **First**, the warfighter already thinks in terms of topography, terrain, maps, etc. This means that by using "human terrain" as a term, and speaking of "culture" in terms of "terrain," soldiers and Marines can talk about culture in a way that makes it sound familiar to the military: graphically representable, quantifiable, and geographically measurable.

Second, there are aspects of culture that one can map. These include demographic features (population density, age distribution, distribution of income, diffusion of services, etc.); social features (kin, ethnic, religious affiliations, etc.); and the location of physical items of symbolic and ideological value to local populations (churches, mosques, shrines, cemeteries, sacred rock formations, etc.). In short, these are all things that one can literally pin on a map.

However, "human terrain" as an applied concept can also harm a warfighter's situational awareness and operational execution. This occurs when "human terrain" is considered more than a metaphor with certain, limited applications. When "human terrain" instead becomes what a Marine thinks culture *is,* and causes him to believe that mapping is *the* way to analyze culture, then "human terrain" has misled a Marine.

Although human terrain maps can communicate information quickly and simply, a two-dimensional representation does not account for dynamic socio-cultural conditions.   Such maps fail to grasp the important interconnections and overlapping relationships among both groups and individuals within the groups.

This leads us to a **definition of human terrain:**

• **Those cultural aspects of the battlespace that, due to their static nature, can be visually represented on a geographic map. Human terrain is static with respect to change over time; rigid with respect to fluid human relationships; and limited to representing behavior in only two dimensions.**

Example: Downside of Taking "Human Terrain" Too Far

A Marine tasked to create a human terrain map of a rural area of southern Morocco might produce a map with blocked out areas indicating distinct "Berber" and "Arab" tribes associated with specific geographic areas. Such a two dimensional map would suggest to the reader that certain areas are "Arab" or "Berber" and that there is little overlap or interaction between the groups. It might also mark out areas of demographic friction, where potential for disturbances would exist because one ethnic group border touched another. However, the realities are far more complex, and such an interpretation would be inadequate.

First, due to intermarriage, some of the so-called Arab households include Berber husbands and wives and vice-versa, resulting in a number of kinship ties which cross-cut ethnic and tribal links and bind the region in relationships of obligation and honor. Second, as a result of labor migration, both to urban areas in Morocco and internationally to France, a large percentage of the male population lives far away from the region.

Example: Downside of Taking "Human Terrain" Too Far, continued.

Research has shown that these migrants tend to cluster in the new cities or countries, in essence re-creating their cultural community of origin far from home.[8] Men in these migrant communities continue to send remittances back home, marry wives from their home communities, and retire back in the region.

These dynamic urban-rural exchanges also cannot be captured on a two dimensional map, which gives the reader a sense that local tribal groups are discrete, disembodied settlements bounded in one area, rather than dynamic fluid communities that span not only national but international boundaries.

Second, a map that color-codes "Arab and "Berber" will tell the operator nothing about whether the color-coded "Berber" feels "Berber," or identifies with that community's cultural and political aims. This is important: during the colonial era in Morocco and Algeria, French authorities wishfully thought Berbers were unlike Arabs in ethnicity and even religion, and were thus more pro-French.[9] French colonists went so far as to attempt to cultivate a different identity among Berbers, so as to use them to lessen the influence of Islam and Arabism in north Africa. Berber difference never became a cultural "rule," however, with some Berbers prominent anti-French nationalists.

Finally, a human terrain map focusing on "Berbers" and "Arabs" cannot account for other criteria for differentiating or uniting people, that in any given time and place might be more prominent than ethnicity. For example, one might argue that in Tunis, Fez, Algiers, or Paris, well-educated, financially comfortable, leftist Berbers and Arabs have more in common with each other (and with well-educated, financially comfortable, leftist French Parisians) than they do with their co-ethnics, back home or in Paris.

In the absence of appropriate social science definitions and explanations guiding the military decision making process, "human terrain"—as either a catch phrase or series of maps—will become inappropriately fused with kinetic military mental models, and result in simplistic understandings of the operational environment. Instead of cultural aspects of the battlespace appearing as dynamic and fluid (see below), it will be too tempting to view humans as "the green order of battle."[10]

Culture: An Operational Definition

In contrast to military definitions, academic definitions of culture are often more theoretically sound. They are, however, just as often too complex, abstract, and ambiguous for military uses. As a token of this problem, literally hundreds of definitions of culture can be found in the academic community.[11] Still, despite the confusing array of definitions, most contemporary scholars agree that culture includes the following key factors:

- Culture is shared.[12]

- Culture underlies our world view: what we perceive and think about the people and events surrounding us, and how we interpret and understand those people and events.[13]

- Culture is interconnected and holistic; each dimension of culture is intimately related to the others.[14]

- Culture is varied—over time, over space, and among individuals.

- Culture is fluid and dynamic; humans are active agents, and not passive recipients.[15]

In academic terms the way that culture shapes and forms our perceptions of the world is referred to as a world view; how people organize their political, economic, and social relationships is referred to as a social structure. If Marines can understand the world views and social structures of a culture, they will be more able to make sense of the resulting actions and choices of people in that AO.

Focusing on an operational basis for understanding culture, in this book we define culture as:

Culture: The shared world view and social structures of a group of people that influence a person's and a group's actions and choices.[16]

This definition emphasizes that cultural beliefs and social structures influence people's actions and choices. By focusing on the *outcome* of cultural beliefs and structures—the actions that people take—the definition gives the Marine an observable behavior (rather than just an unobservable belief system), which can be incorporated into operational planning.

The word "culture" not only refers to the meanings and beliefs that people hold, but sometimes to a specific group. To distinguish this meaning, we also need to define a culture group.

Culture Group: A group of people whose common world view unites them in a system of social structures and shared behaviors.

Again, it is important to recall key terms just presented.

World View: The way that culture shapes and forms our perceptions of the world.

Social Structure: How people organize their political, economic, and social relationships

From these basic foundational definitions of "culture" and "culture group," one can proceed to grasp key aspects about the nature of both. In order to plan and operate successfully according to sound cultural principles, Marines need to recognize and contemplate the implications of the following characteristics of the phenomenon called "culture." We will now examine each of these characteristics in detail.

Characteristic One: Culture is shared. In a sense culture serves as a shared "language" among people. Those people who belong to a cultural group understand the language—the social symbols (whether it is a star upon a collar or a Christmas tree), the meanings attached to certain behaviors (for example what a salute communicates about status and rank) and the importance of relationships to people in the group (the authority of a Colonel versus the friendship of a fellow boot camp recruit; we return to symbols and their social meaning in Chapter Seven).

Included in this concept of culture is the notion that, like language, culture is learned. As soon as a child is born, he or she is taught not only what behaviors are appropriate for that group, but also the meaning of those interactions with others. People can learn about and become members of other cultural groups at any time in their lives. Marines, for example, generally are socialized into the Corps as adults. As with the Marine Corps, this learning process is both explicit and implicit.

Because culture is shared, *individual* beliefs, ideals and meanings that are unique to a specific person do not constitute culture. Sometimes we mistake the unique characteristics of a *person* from a foreign culture to be representative of that *group as a whole*. In working with people from different AOs, Marines must distinguish between those behaviors and attitudes that are shared by most people in the culture, and those that reflect individual preferences.

Characteristic Two: Culture underlies our world view: what we perceive and think about the people and events surrounding us, and how we understand those people and events. As such, our way of thinking about the world, of organizing and evaluating our experiences is framed by our cultural background.

However, although world view influences the way a person thinks about his or her world, it does not determine a person's actions. Culture is not a thing that exists somewhere outside of people: Individuals have a vote, and *they interact with received ideas through the prism of their experiences*.

Example: Behavior as the Product of Adjustment to Reality

In many areas, Marines encounter a "culture of theft"—a cultural attitude that seems to condone stealing as an acceptable social behavior. But what they really see is peoples' reactions to specific conditions, such as the scarcity of basic commodities, the breakdown of normal commercial methods of exchange and currency, and low levels of inter-personal trust. Each of these factors in turn results from conditions—such as government inability to provide economic and food staples in sufficient quantities, based on a state-run system of production which underpays employees.

Example: Behavior as the Product of Adjustment to Reality; continued.

These conditions are produced by market forces regionally within the country, and also globally, as the price for fuel, transportation, etc., increases, the national balance of trade becomes negative and so on.

So, global economic <u>forces</u> interact with political/economic <u>conditions</u> in a country and local region to create specific <u>circumstances</u> to which <u>people react</u> in a way that manifests in a "<u>culture</u> of theft" which Marines see.

As this example shows, to be useful in a military sense, it must be understood that "culture" or cultural factors alone do not determine how people act. Instead, **forces which humans encounter create physical conditions** and events. **People then respond to these conditions and events**, based on their culturally determined world view of possible acceptable ways to act. **Marines thus see the product of physical conditions interacting with culture to influence behavior and social conditions.**

Characteristic Three: Culture is interconnected and holistic; each dimension of culture is intimately related to the others. This means two things. First, culture is not an amorphous black box, but composed of interconnecting beliefs, behaviors, social structures and relationships. These various aspects or dimensions of culture must be analyzed and understood in relationship to each other. Changes in one dimension, such as the economy, can have profound effects on another dimension, such as political structure and power.

Secondly, culture is holistic; each of the parts is intimately related to the others. Changes in one part of a culture (its economic system for example) may thus have profound effects on another part (the political structures).

Culture is a phenomenon greater than and different from the sum of its parts—just as salt (Sodium Chloride, NaCl) is greater than and different from Na mixed with Cl. Therefore, culture cannot be a "check in the box" list of items or behaviors, but must be treated as an integrated way of looking at the world.

Characteristic Four: Culture is varied—over time, over space, and among individuals. Culture is not a homogeneous system that everyone follows without thinking. Within a culture group there will be much variation in terms of what people actually believe and do. Not all people within a culture group will have the same cultural knowledge or experiences.

Along with this internal variation, individuals themselves are culturally varied. They belong to many cultural groups and hold ties across social and cultural groups. As such, an individual's attitudes and choices reflect personal choices about competing cultural ideologies and memberships. Put differently, **there is not just one culture per person. In fact, one person can be a member of many cultures or sub-cultures.** There are thus multiple pieces in the jigsaw puzzle of the "culture" a person presents to others. Each of these pieces influences how a person processes and acts in the world.[17]

Example: I.M. Marine

Consider a U.S. Marine, whom we will call "I.M. Marine." If he is from a family of Hindi-speaking immigrants, he has been partially socialized through his upbringing in an Indian family. If that family is Muslim, that socializing element has also been at play. Growing up in Tennessee and attending a very large public high school have also played into his cultural formation, as has the socialization process of Boot Camp, the School of Infantry, and fifteen years of a career on the West Coast, for example.

In the table on the next page, we see multiple possibilities for one's family, language, national, etc., memberships. In this case, "I.M. Marine" possesses the cultural affiliations in bold type.

As we see here, "I.M. Marine" has many cultural jigsaw pieces to his personality puzzle. It is by understanding his particular cultural complexion, and his own choices and preferences, that we can understand him. More broadly, the case of "I.M. Marine" shows us that while we should not confuse culture with individual personality, "for any individual, culture always comes in the plural."[18]

Table 1: I. M. Marine: Multiple Cultural Memberships				
Family	1 Generation Immigrant	Native	2 Gen Immig	Alien
Language	French	Arabic	**Hindu**	German
Nation	Egypt	**India**	Norway	Greece
Religion	Christian	Buddhist	Jewish	**Muslim**
Gender	Female		Male	
Region	Tennessee	California	Massachusetts	Florida
Schooling	Military	**Public HS**	Private	Home
Job	Doctor	Bus Driver	Lawyer	**Marine**
Specialty	**Infantry**	Intel	Motor T	Comm
Dwelling	NC	Florida	**S California**	N VA

Cultural influences are not only found in traditional organizations, such as the family, ethnic groups, religious communities, or tribes. Modern workplaces also have their own cultures, as do schools, activist groups (such as Greenpeace), and even subversive groups (such as the Mafia).

Example: Cultural Influences Beyond Traditional Categories

A unique international culture has emerged among people who have attended elite universities and then pursued a career in the foreign service of a country. These people frequently speak several languages, have friends from many countries and often have spent more time away from their home country, than in it. This could be referred to as the "diplomatic elite culture" of a country, and in general it crosses national borders. As another example, The five services of the U.S. military all have definable "corporate cultures," consisting of self-image, ways of doing business, material means, rules, histories, and mental models.

Further, each service posesses identifiable communities, such as intelligence officers, aviators, infantrymen, special operators, etc. In these cases, a recognizable culture has emerged from institutions or among a group of people with common experiences and aspirations. Though not based on "religion," language, ethnicity, or elements commonly assumed to constitute culture, they have indeed culturally influenced people.

These examples also illustrate that culture is by no means homogeneous. **Generalizing about mass populations does not help the warfighter,** because such generalizations miss the important multilayered and multifaceted nature of human existence. By reducing individuals to a single tribal group or ethnic group we fail to grasp the important interconnections between individuals and groups—information that could be critical in understanding insurgent networks, the movement of illegal goods, or ties of power and alliance in a region.

Characteristic Five: Culture is fluid and dynamic; humans are active agents, and not passive recipients. Since culture is the product of *human* thought and behavior, each person and each generation redefines and recreates the culture of their group. Through their responses to and attitudes towards events, people create culture, and not the other way around. As a result, each individual and each generation will redefine and recreate their culture.

Likewise, although people may share a certain core set of cultural beliefs and attitudes, that does not mean every person agrees with or behaves consistently according to those ideals and values.

Warfighters should not assume cultures remain constant over time. Cultures are made and remade as culture-influencing parameters squeeze and remold societies. For example, in 2007, the culture of the middle-class, educated Muslim Iraqi from Baghdad is likely notably different in certain key operationally-relevant respects from what it was in 1988—not to mention what it was like in the days of T.E. Lawrence or Gertrude Bell.

For the warfighter, the cultural dynamics that matter change from week to week, neighborhood to neighborhood, and mission to mission. **When Marines think about culture in a way to use it on the battlefields of today and tomorrow, it is important that experiences in previous operating environments not limit their thinking.**

Example: Iraqi Cultural Changes Since 2003

There are various instances of change in behavior, attitude, and demographics since the beginning of OIF. Some are very small. Whereas before 2003, the thumbs-up sign was not widespread among Iraqis and had an obscene connotation, repeated use of it by American personnel in a positive way has led some Iraqis, particularly younger ones, to use it in a positive way when communicating with coalition personnel. Likewise, certain military or technical terms which had been used by Iraqis in Arabic have now come to be used in English; conversely, British usages (such as "lorry") are being replaced by American usages (such as "truck").

More fundamental changes have resulted from rising inter-communal violence. While use of the female headscarf (*hijab*) was noticeable in Iraq before 2003, the rise of Islamist violence and intimidation has led to a much greater incidence of Iraqi Muslim—and sometimes Christian—females wearing *hijab*, to either avoid intimidation and abduction, or assert a belonging (even if fictional). Similarly, Sunni and Shi'i men and women have been changing their names to seem less Sunni ('Umar, 'Uthman, etc.), or Shi'i ('Ali, Haydar, Zaynab, etc.), thus reducing the utility to coalition forces of names as ethno-religious markers.[19]

Finally, the ethno-sectarian map of Iraq has changed, rendering earlier "human terrain" maps obsolete. In Kirkuk, the proportion of Kurds to Arabs, and their locations, have been changing, as Kurdish rulers in the city pressure Arabs to leave and facilitate Kurdish return. In Baghdad, mixed Sunni-Shi'ite neighborhoods are re-aligning, such that Sunnis are moving out of Shi'ite areas, Shi'ites are leaving Sunni areas, and Christians are emigrating.

Therefore, Iraqi culture has changed over the course of OIF: in terms of body language, verbal language, dress, and demographics.

From "Culture" to "Operational Culture"

It is tempting for Marines to accept descriptions of culture that seem to crack into the "mindset" of people. Such approaches portray culture as a special potion, or silver bullet that gives Marines secret knowledge. However, simplistic descriptions of culture that explain a person's behavior based upon the notion that he is from a "Confucian civilization," or has an "Arab mind" are of little use to the operator. They can result in operational plans hinging on static assumptions about culture as a code to be cracked, or a behavior-predictive blueprint, akin to the doctrine of the other side's "green order of battle."

In this text we argue that culture shapes our world view; it is composed of a pattern of relationships and structures and is varied and dynamic; It is not a check-the-block factor that can be reduced to a map and predicted with scientific certainty. Yet despite its complexity, culture can be understood and included in operational planning, training, and execution. When that occurs, the result is "operational culture," which we define here.

In Chapter One, we began our discussion of operational culture by calling it **"those aspects of culture that influence the outcome of a military operation; conversely, the military actions that influence the culture of an area of operations."** This phrasing recognizes that many items not within the civilian or academic understanding of culture are indeed of concern to the warfighter. This brings us to a fuller definition.

Operational Culture: Governed by a particular operation's goals and material assets, as well as the functions of personnel associated with a particular operation, Operational Culture consists of

• Operationally relevant behavior, relationships and perceptions of indigenous security forces against or with whom Marines operate; civilian populations among whom Marines operate; indigenous communities or groups whom Marines wish to influence; international partners in coalition operations.

• Dimensions influencing operationally-relevant behavior, conduct, and attitudes. These Operational Culture Dimensions involve the physical environment, the economy of a culture, social structures, political structures, and the beliefs and symbols of a culture group. These dimensions emerge from three major models of cultural analysis: the ecological, social structure, and symbolic models.

• Historical trends that influence the interaction among those cultural dimensions

• Capability to successfully plan and execute across the operational spectrum, including humanitarian assistance and disaster relief; pre-hostility; shaping operations; successive campaign phases; and post hostilities, to include reconstruction and stabilization, as well as peace making/keeping.

For the Marine, Operational Culture is not simply a thing, or feature of the battlespace, outside him or her. Rather, it is a continual process of individual and collective *learning* about contemporary and future operations:

Operational Culture Learning: Learning about cultural concepts and dimensions, across the spectrum of pre-deployment training, professional education and formal schools, and individual professional development.

• In the predeployment phase, scaled to rank and billet and focused on the AO as aligned with mission goals: **study** of a specific AO's culture, to include expressed behaviors and attitudes, as well as the interaction among Operational Culture Dimensions which produce these behaviors and attitudes; **training** in operation- as well as billet-focused language domains; attendance of command-provided distance learning, function-focused face-to-face seminars, field exercises, and situational training; and **self-study**, to ensure learning in the cognitive, psychomotor and affective domains.

• In PME, reflecting the responsibilities of Marines at the completion of each stage of PME: **study** of the fundamental concepts and Dimensions of culture *in generica*; **development** of skills to function successfully in various geographical and diverse human environments; **examination** of human, print, and electronic resources for learning about operational culture; **exploration** of the historical role of the Five Operational Culture Dimensions in the battlespace through study of past areas of operation; and introduction to the **application** of concepts and skills to the current operating environment.

- In the career continuum, appropriate to MOS, phase of career, and leadership responsibilities: **study** of emerging operating environments; **maintenance** of capabilities with respect to regions of past or likely future areas of operation; **monitoring** service- and DOD-provided training and educational resources for culture learning; **fostering** within units and commands continued culture learning and an atmosphere supportive of individual Marines' study of foreign cultures; **recording** culture-relevant observations about areas of deployment.

Likewise, one who recognizes the nature and significance of operational culture, and engages in continuous operational culture learning, embraces the identity of a "culture operator."

Culture Operator: A warfighter who engages in military functions at the tactical, operational, and strategic level within his AO through **continually re-reading** the changing cultural and human aspects of the battlespace as they impact military operations; by **tracking the dynamic interaction** among the Operational Culture Dimensions of environment, economy, social structure, political structure, and beliefs and symbols; and by **considering the impact of Marine operations** as a new physical condition of human existence for indigenous people in the AO, influencing local behaviors and attitudes.

Finally, whenever possible, a "culture operator" engages in "culture operations."

Culture Operations: Utilizing the full complement of kinetic and non-kinetic effects, "Culture Operations"

- Include cultural and human factors in operational planning, while considering possible responses of people in an AO to Marine actions during operations, and evaluating cultural outcomes of tactical and operational measures. In culture operations, culture operators weigh cultural outcomes of tactical and operational measures against mission objectives, and develop innovative courses of action allowing commanders previously unrecognized opportunities.

• Culture Operations, based on contextually appropriate knowledge of the local culture group, allow the Marine to frame Marine planning and execution in order to create conditions facilitating conduct by indigenous people commensurate with tactical or operational goals.   Marines can thus select most appropriate operational methods for short-, medium-, and long-term purposes; whilst mitigating local peoples' objections to necessary actions which contravene local cultural norms.

Planning and execution of Culture Operations must recognize that Marine actions are merely one among many factors influencing human conditions in the battlespace.   Marine actions rarely control the many cultural and other factors that interact with each other across the phases of engagement in the AO.

Furthermore, while Marine actions might *influence* behaviors or mechanisms producing behaviors, they are unlikely to *craft* or *determine* people's behaviors.   Likewise, Marine actions cannot *determine* how people respond to Marine *influences* on their conditions of existence.

These definitions and related considerations indicate what constitutes operational culture, what understanding it can do, and what are its limitations.   We have here also articulated what it means to be a Marine who integrates culture in professional learning, pre-deployment training, and individual growth.   Furthermore, these definitions, as well as earlier chapters, imply that to be a Marine capable of operations in hybrid environments, one *must* be a culture operator.

Conclusion

Beyond definitions, Operational Culture Learning, or Culture Operations, deal with the concrete world, and human features which Marines encounter while deployed.   To get to the substance of this concrete world, a Culture Operator must remember the three anthropological models discussed in Chapter One which possess utility in military operations. To review, these were the **ecological model** concerning the relationship between cultures and the physical environment; the **social structure model** examining the influence of social structures on peoples' status and power; and the **symbolic model** studying a group's beliefs, symbols and rituals.

As we saw, these three models produce the five dimensions of operational culture: the **physical** environment, the **economic** system, the **social** structure, the **political** structure, and **beliefs** and symbols. In Part II of this book, we turn to an examination of each of these dimensions in detail, to include their components. As we have suggested up until now, the interaction among these dimensions, and the intermingled nature of their effects on human societies, will be a central matter of concern.

Notes

[1] Maxie McFarland, "Military Cultural Education," *Military Review* March-April 2005, 62-63.

[2] Margaret Mead and Rhoda Metraux, *The Study of Culture at a Distance* Berghahn Books, 2000 [1953]), 22.

[3] Edward Hall, quoted in Peace Corps Information Collection and Exchange, *Culture Matters: The Peace Corps Cross-Cultural Workbook* (Washington, DC: United States Government Printing Office, n.d.), 18.

[4] Ibid., 15.

[5] William D. Wunderle, *Through the Lens of Cultural Awareness: A Primer for US Armed Forces Deploying to Arab and Middle Eastern Countries* (Ft Leavenworth, KS: Combat Studies Inst. Press, 2006), 9.

[6] United States Army John F. Kennedy Special Warfare Center and School, *Special Operations and International Studies: Political-Military Analysis Handbook* (Fort Bragg, NC, 2004), 13/1.

[7] The term "Human Terrain" appears to have been first used in a militarily meaningful fashion by Ralph Peters, in "The Human Terrain of Urban Operations," *Parameters* Spring 2000, 4-12: http://carlisle-www.army.mil/usawc/Parameters/00spring/peters.htm.

[8] Janet Abu Lughod, "Migrant Adjustment to City Life: The Egyptian Case," in Ibrahim Saad and N.S. Hopkins, eds., *Arab Society in Transition* (Cairo: American University in Cairo Press, 1977), 391-405.

[9] See Patricia Lorcin, *Imperial Identities: Stereotyping, Prejudice and Race in Colonial Algeria* (London: I.B. Tauris, 1999).

[10] Author's interview with Brigade Combat Team Deputy Commanding Officer, Baghdad, January 2008. Our critical comments here focus on misuses of the

term and concept of human terrain. This is very different from "Human Terrain System." This is (as of 2008) a joint program in its initial stages of development, sponsored by U.S. Army's Training and Doctrine Command. Human Terrain System provides operational ground force units at the battalion-and-higher level with social scientists, field researchers, and cultural analysts with military backgrounds (Human Terrain Teams), in order to assist commanders and staffs in every stage of battlespace evaluation, mission analysis, and course of action development, in the specifically non-lethal lines of operation. For more on this see "'Academic Embeds': Scholars Advise Troops Abroad," *Talk of the Nation* (National Public Radio), 9 October 2007: http://www.npr.org/templates/story/story.php?storyId=15124054. For an example of Human Terrain Teams deployed, see "Policing a Whirlwind," *The Economist* 13 December 2007: http://www.economist.com/world/asia/displaystory.cfm?story_id=10286219 and David Rhode, "Anthropologists; Scholars embedded in Afghanistan" *The International Herald Tribune,* (October 6, 2007),3.

[11] For example, Kroeber and Kluckhohn collected more than 160 definitions of culture in their classic work: A. L. Kroeber and C. Kluckhohn, *Culture: A Critical Review of Concepts and Definitions* (Visalia CA: Vintage Press, 1954).

[12] See Gary Ferraro, *Cultural Anthropology: An Applied Perspective.* 5th edition. (Belmont CA: Thomson Wadsworth, 2004) and Bailey Garrick and James Peoples, *Introduction to Cultural Anthropology* (Belmont CA: Wadsworth, 1999).

[13] Clifford Geertz, *Interpretation of Cultures* (N.Y.: Basic Books Classics, 2000).

[14] See Ruth Benedict, *Patterns of Culture* (N.Y.: Mariner Books, 2006); Bailey Garrick and James Peoples, *Introduction to Cultural Anthropology.*

[15] See Jerry Moore, *Visions of Culture: An Introduction to Anthropological Theories and Theorists* (Lanham, MD: Alta Mira Press, 1997).

[16] This definition, which includes the concept of social structures, leans heavily on a British Social Anthropological perspective. In the U.S., American symbolic and structuralist analyses have become the norm in recent years, led by anthropologists such as Clifford Geertz and Marshall Sahlins. See, for example, Geertz's *Interpretation of Cultures* and Sahlins' *Islands of History* (Chicago: University of Chicago Press, 1987). Current American structuralist approaches are strongly informed by post-modernist debates focusing on the nature of knowledge and its meaning, particularly in light of the role of the ethnographer's culture in understanding the "Other." The authors acknowledge the extent and importance of current American structuralist thinking—in particular its contribution to the notions of identity, meaning, process and symbolic communication. And throughout the text, we emphasize the dynamic and symbolic nature of culture and rely heavily on American structuralist approaches. However, the definition of culture in this book focuses on the concept of social structure since the British Social Anthropological approach is more concrete, and its applications more clear for Marines operating in conflict zones and other deployed areas.

[17] Jean.-Claude Usunier, *Marketing Across Cultures* (New York: Prentice Hall, 1996).

[18] Kevin Avruch, *Culture and Conflict Resolution* (Washington, DC: USIP Press,

1998), 15.

[19] Deborah Amos, "Iraqi Women Fight for Voice in New Iraq," *Morning Edition* (National Public Radio), 3 July 2003: http://www.npr.org/templates/story/story.php?storyId=1319135; Jamie Tarabay, "What's in a Name? In Iraq, Life or Death," *All Things Considered* (National Public Radio), 9 May 2006: http://www.npr.org/templates/story/story.php?storyId=5394090.

Part II

Five Operational Culture Dimensions for Planning and Execution

Gaining a complete and accurate understanding of a culture requires spending years living in the region, learning the language and interacting with the local people. These are things that the majority of Marines cannot do—and focus on warfighting functions at the same time. However, what a Marine can and should come to recognize is that despite their seeming diversity, all cultures are organized according to a predictable set of categories or dimensions. Although the specific details will vary (often drastically) from one culture group to the next, these dimensions can be found in any culture anywhere in the world. For reference purposes we summarize these five dimensions here.

The Five Operational Culture Dimensions of the Battlespace

1. Environment. All cultures have developed a unique interdependent relationship with their **physical environment.**

2. Economy. All cultures have a specific system for obtaining, producing and distributing the items that people need or want to survive in their society. This system (which does not necessarily require money or banks) is called the **economy of a culture.**

3. Social Structure. All cultures assign people different roles, status and power within the group. The way that people organize themselves and distribute power and status is called their **social structure.**

4. Political Structure. All cultures have a system that determines who leads the group, and how they make decisions about its welfare. How a group is ruled (and it may not be by a specific person or set of people) is referred to as the **political structure** of a culture.

5. Belief Systems. All cultures have a shared set of **beliefs and symbols** that unite the group.

In Part II, we will discuss the Five Operational Culture Dimensions of the battlespace that Marines can examine and incorporate into planning and execution. Chapter Three will focus on the relationships among the environment, culture and military operations. Chapter Four examines three forms of economic relationships from a cultural and military perspective. Chapter Five looks at social structures of different cultures, focusing on six factors that affect one's status, role and power. Chapter Six analyzes power and authority in political systems, emphasizing cultural concepts of leadership. Finally, Chapter Seven discusses the impact of cultural beliefs on operations.

Chapter 3

Dimension I
The Physical Environment

This chapter begins to explain the Five Operational Culture Dimensions in detail, by focusing on the physical environment of a culture. Each of the following chapters will then discuss the remaining four dimensions. Part III will then provide case studies from ongoing or historical conflict areas, demonstrating the dynamic interaction among cultural dimensions, as well as the implications for Marine operations and planning. We will then examine the integration of culture into Marine learning.

As seen in subsequent chapters, each Operational Culture Dimension sets the stage for, or exists in recursive relationship with, other Dimensions. Put simply, human phenomena—like other elements in the battlespace—are intermingled, and artificially separating them out in operational planning might provide a deceptively tidy sight picture. In this and following chapters, therefore, we will refer to multiple dimensions allowing the reader to begin to trace out the linkages on his/her own.

How Culture Groups Relate to their Environments

Since ancient times when man first depended on the environment to hunt and gather food, human groups have created unique relationships with the physical geography of a region. The availability of food, water, and material for shelter has determined the location of villages and towns. Nomadic herding routes have traditionally followed the availability of food and water for animals. And even today, most roads and the cities along them are built on ancient trade routes that trace the easiest passage along rivers and through mountain passes.

In preparing an operational plan that includes cultural factors, Marines need to understand the close relationship between a local community and its environment. Fundamentally, Marines need to determine what features of the local physical environment are used by people in their daily effort for survival. This is because how Marines use the physical environment in the course of operations influences the ability of indigenous humans in that AO to survive. Likewise, how people already use the physical environment will inform the range of options available to Marine forces entering an area.

Example: Operational Impact on Physical Environment

Cedar and Birch trees are prevalent along the Northwest Coast of the U.S. and Canada, ranging from Washington to Alaska. Indigenous Native Americans such as the Tlingit, Kwaikutl, Haida and Susquatch in these areas have traditionally dried the bark from these trees, and then beaten it into a material from which they construct baskets, mats, and bent-wood boxes. These products are highly prized for their beauty and workmanship and command high prices in art and tourist stores across the U.S.

The physical environment's trees, therefore are used by indigenous people to create income upon which the community depends. If Marine forces were to choose a section of the Washington or Alaskan coastline for air-naval gunfire and amphibious exercises, they might either clear the coast of the trees, fire upon them, or at the least, declare it a zone closed to civilian traffic. In this case, Marine operations would impact the physical environment, with negative consequences for the life, and material survival, of people in the local culture group. Operations would also create potentially hostile attitudes towards the Marine presence.

Physical features of an environment often possess a significance that goes far beyond their mere physicality, however. Sometimes people in culture groups attach to an item or feature of the physical environment an importance different from or greater than its physical value; or, they might use that physical thing in a way different from its obvious physical usefulness. As important as *ma-*

terial survival, the *symbolic* importance of physical features of an AO is a core matter of concern to Marines. We will encounter this topic again in Chapter Seven.

Example: Physical Environment and Symbolic Value

Mats, baskets, and boxes constructed from cedar and birch bark in the Pacific Northwest have a local cultural significance beyond economic exchange. Traditionally, local Native American tribal leaders have communicated status through demonstrations of hospitality, known as potlatches, or gatherings where food and gifts are distributed. High-status gifts which tribal leaders can bestow during potlatches have included boxes, and in particular, baskets, constructed from cedar or birch. Logging in areas used by local Native American groups would thus negatively affect the availability of cedar, birch, and the products derived from them. More important symbolically, logging would make it more difficult for tribal leaders to demonstrate patronage and status in traditional terms.[1]

This would drive a shift in the groups in society who have status, and how they demonstrate it. If amphibious exercises also impacted the availability of these physical features with symbolic significance, they might influence the attitude of traditional local group leaders towards the Marine presence.

Operationally Relevant Features of the Physical Environment

Water. Perhaps no other physical resource is more precious than water. Since ancient times, man has fought to the death over access to and control of water. In much of the developing world today, getting access to water (and in particular, safe, drinkable water) is a serious issue. In some arid areas, such as the Arabian Peninsula and the Horn of Africa, local peoples may spend up to half of their day hauling water from wells or watering holes to their homes and fields.

Because water is so precious, many societies have intricate and often unspoken rules that regulate the use of water. For example historically, across the Arabian Peninsula different tribes controlled the watering holes or oases scattered throughout the desert. For centuries in India, a specific caste has held the responsibility for bearing and selling water. And in the U.S. today, carefully detailed laws regulate how and when water is distributed for irrigation of crops in eastern Washington and Oregon.

Clearly, water is a major source of conflict around the world.[2] Conflict over water often has a technical or access-related aspect to it, particularly to external observers. Beyond the technical aspects, however, cultural and political elements are often at play, involving notions of access, control, and use, as related to ideas of sovereignty, historical rights, and past feuds—items we shall discuss later, in Chapters Six and Seven, when we examine political structures and beliefs.

Example: Water, Access, and Status

During the 1980s-90s, Turkey was in conflict with Syria and Iraq over the Tigris and Euphrates Rivers. Turkey had planned to dam the rivers since the late 1970s, and in the 1980s began to do so, in order to increase hydroelectric power and economically develop southeastern Anatolia. This "South-Eastern Anatolia Project" sought to effect better standards of life, provide investment and employment opportunities in a long-neglected region, increase government control of the area, and lessen the local Kurdish population's attraction to separatist terrorism. However, damming the Tigris and Euphrates reduced the water flow downstream to already water-strapped northern Syria and Iraq. These were the technical aspects of the water dispute.

Perhaps of more enduring significance to relations among these countries were the cultural issues over-laying the technical matters. As a large regional state with a large army and industrial potential, Turks at a popular level thought they deserved to use waters which were sovereign Turkish resources. As the headwaters were located in Turkey, the water was "their" patrimony. By contrast, Syrians and Iraqis considered Turkey a ham-fisted interloper into the region—different ethnically, divergent in interpretations of religion, and the inheritor of a "bad attitude" based on the previous Ottoman empire's imperial pretensions.

As a response to Turkey's water use, therefore, Iraq and Syria both supported separatist Turkish Kurdish movements with training bases, safe passage, and weapons. This in turn evoked nationalist ire among Turks in and out of government, who could point to "Arabs" stabbing them in the back again, as they had in WWI during the "Arab Revolt" in the Hijaz.

Example: Water, Access, and Status; continued.

Thus, opposition from the Arab states to Turkey about the physical asset of water was often equally related to cultural notions of empowerment, control, sovereignty, and historical memory (see Chapter Seven), thus rendering sensible technical solutions at times irrelevant to the "real" issues at stake.[3]

Culture Operator's Questions: Water

Marines therefore need to ask certain questions about water as a physical resource in the AO:

- What are the cultural rules about water's use?

- What roles are expected of Marine personnel with respect to water use and provision?

- What is the relationship between water use and ritual?

- What is the symbolic significance of water?

- Who customarily has what functions with respect to water?

- Who, in the AO, has customarily controlled access to water, and how have they used that for power, influence, etc.?

- What is the scarcity of water in relation to intensity of use?

- What operational considerations are influenced by water, or override cultural aspects of water as a physical resource?

Land. Land, like water, has been the source of conflict between people for thousands of years. From a cultural perspective, land is much more than the physical geography. For many cultural groups, certain places hold a symbolic meaning that is significantly greater than the simple physical features of terrain. For the Navajo Indians of the American Southwest, for example, many of the

strange rock formations on their reservation have stories and legends connecting the Navajo to their mythical beginnings. These
rocks serve as centers of sacred ceremonies and are considered
the heart of their culture. Likewise, in many parts of Central Asia,
gravesites of wandering mystical preachers (dervishes) who
brought Islam to the region between the eleventh and eighteenth
century have become shrines, to which people make pilgrimage to
pray and study. The land, in these instances, has become associated with holy personages, and is assumed to have sacred proper-
ties.[4]

Sometimes the symbolic significance of land means that some people are allowed to be on a certain patch of ground, whereas others are not. Arabia, and the holy cities of Mecca and Medina in
particular, have traditionally been off-limits to non-Muslims. The
influx of U.S. personnel into Saudi Arabia after 1990 has caused
cultural animosity among some Saudis. Likewise, in traditional
Jewish culture, the Temple Mount in Jerusalem has been off-limits
to Jews, since its holiness requires its access to be limited to the
High Priest of the (defunct) Temple Cult. Marines must therefore
consider the symbolic importance both of their presence on particular patches of land, and their conduct on that land. For example, in the opening phases of OIF, coalition forces passing through
Kut and Amarra restored the British military cemeteries in these locations. This was interpreted by local Iraqis as Americans following in British footsteps to assume the role of imperial conquerors.
The cemeteries were in some instances subsequently re-desecrated.

Land may not only have symbolic meanings that outsiders do not
recognize, but the use and ownership of land may be viewed differently by other cultures. For certain groups around the world, the
idea of landownership is foreign to their cultural world view. For
the Mongolians, for example, land is communal. Out in the Gobi
Desert there are few fences, no walls and no "keep out" signs. Instead the land is shared and used as needed by each household.
Following their herds of animals in search of food, Mongolian families move their *gers* (or round tent homes) across the desert. They
put up their tents where it is convenient, and move on when the
animals need fresh grazing pastures.

By contrast, in some socialist, communist, and even monarchical countries, land is owned by no one but the state, though use is (theoretically) open to all. Therefore, understandings of ownership, use, and legal status of land are culturally coded, with potential operational ramifications.

Example: Land Ownership, Land Use, and Cultural Conflict

Different notions of ownership and proper use can cause conflicts between cultural groups. In early 20th-century Palestine, European Jewish settlers purchased land titles from absentee landowners, often living in what is today Lebanon. According to European ideas of landownership, those who purchased title to a parcel of land owned that land, and could determine who farmed on it and how.

By contrast, the Palestinian peasants living on and farming the land did not adhere to a system of individual landownership. Rather, they had been communally farming lands. "Title"—a new-fangled practice from the far-away imperial capital in Istanbul—meant little to them. When the new Jewish owners arrived and began occupying the land, they assumed that the Arab peasants were invading their rightfully owned space. These two different understandings of ownership led to disagreements, bloodshed, and an aggravation of Arab-Jewish relations.

Likewise, among the Galilee Bedouin, it had been common to reap from fields cultivated by others. Picking a reasonable, limited amount of fruits and vegetables from the margins of fields was considered acceptable in the local Bedouin culture. Zionists, however, considered it brigandage or out-and-out attacks on their lands, with the (political and ethnic) purpose of destroying the new Jewish agricultural settlements. These different cultural practices with respect to land, and the misunderstanding they caused, made political and armed conflict worse between Arabs and Jews.[5]

The culturally-coded nature of approaches to land also holds true for understandings of international boundaries and borders. While these might appear important on a political map, in many regions, these are mere products of grease pencils on a canvas, not at all matching local social or economic reality. One Marine attached to Combined Joint Task Force Horn of Africa found in 2004 that Ugandans and Kenyans saw very little relevance in the British-drawn international border between them. One day, a Kenyan citizen, living in Kenya, asked the Marine, "How are things [over there] in Kenya?" He felt that he was in the land of Uganda as he saw it; he felt more connected to the kin group in Uganda, and did not care where Kenya started.[6]

Thus, an AO's political map is usually broken down further into administrative subregions, often at odds with the tactical map, which often follows a topographical logic. First, a Marine should grasp an area's principles of division, as well as the relationship between these dividing lines and access to resources—tangible *and* symbolic. Then, one can understand how people relate to places in their attitudes and aspirations.

Of course, political powers can use boundaries creatively in spite of cultural realities. This is because geography unites and divides people, resources, and power, as it breaks national states into smaller units, from province to neighborhood. Local, regional, and national powers take advantage of specific interests and internal/external disputes related to geography and resource access.[7] This creates bizarre cartographic contortions, which might cause complications for Marine operations.

Example: "National" Borders and Cultural Realities

Soviet leaders in Moscow regularly gerrymandered internal republics' administrative borders in ways that had nothing to do either with local linguistic or cultural groups' understandings of what patch of ground belonged to which group, or with the local geography of economic exchange. Soviet leaders did this to counter nationalist sentiment in an empire that had no ethno-linguistic commonality. Gerrymandering was particularly widespread in Central Asia, as well as the Baltics. The post-Soviet republics of Central Asia have inherited these maps, and have not changed them, causing some quite strange relationships between peoples' modes of survival and the administrative requirements of "national" territories.

Parts of Central Asian states today are surrounded by other states' territories. Farmers in the Uzbek enclave of Shakhimardan, for example, which is fully encircled by Kyrgyzstan, now have to pass through Kyrgyzstan to get to the rest of their country. As a result, they must pay fees and maintain documents, in order to do business over borders that have no meaning to

Example: "National" Borders and Cultural Realities; continued.

them. This encourages lying, bribery to corrupt border officials, increased prices due to customs duties, and a shadow area conducive to organized crime and terrorist transit.

Similar Soviet practices in Armenia and Azerbaijan caused a shooting war between those two countries in 1991-3 over Nagorno-Karabakh, and a stalemate since then that is perfect fodder for extreme politicians in both countries, as well as for international terrorists from surrounding areas.

Be it in undertaking kinetic operations in a new area, setting up firm bases, or conducting civil-military and reconstruction operations, Marines need to understand both the value that local peoples place on land they use, and local perspectives on "what is mine, what is yours."

Culture Operator's Questions: Land

- What are the symbolic meanings of certain sub-districts in the AO, and how do groups within the AO view this symbolism differently?

- What are particular land formations that are visually striking, with local significance?

- What land in the AO is/is not appropriate for certain groups of people to use?

- Who, locally, has legitimate ability to determine outsiders' access to land?

- What are the local conventions of private, communal, and state ownership/use of land?

- What is the relationship between the political map's national/regional boundaries, and what people living in the AO see as the boundaries that matter, in political, economic, genealogical, and security terms?

- What are the geographic area's principles of division, and relationship between dividing lines and access to both tangible and symbolic resources?

Food. For much of the developing world, obtaining and preparing food is a highly labor intensive task. Most importantly, because of the scarcity of food in some regions, cultures may eat creatures and plants that most Americans would not even consider to be food.

In pastoral societies, such as the Masai of East Africa, food comes from the herds of animals, which are the lifeline of the group. In the case of the Masai, the precious cows are not killed, but bled and milked daily to provide a nourishing drink made from a mixture of blood and milk. In China and parts of Southeast Asia, rats, snakes, scorpions and even grasshoppers form the basis of local diets. And in Poland, a popular pastime is gathering berries and mushrooms from the forests—a practice that often provided needed food during Communist times when stores had empty shelves and thousands went hungry. Marines need to recognize the many items that may serve as the basis of local diets. In their operational plans they should include the impact that their actions will have upon the ability of local peoples to obtain this food.

Food also often serves important ritual and symbolic purposes in other cultures. Some foods may be forbidden, such as meat from a cow in Hindu societies or pork in Muslim and Jewish communities. Other foods may be essential in ritual ceremonies such as the kava drink in Polynesia. And most societies have foods that are served only on special occasions or to special guests. Marines need to be aware of local food taboos, particularly when providing food to a community. They should also be prepared to gracefully accept certain dishes (such as sheep's heads and intestines in North Africa) in order not to offend the host, who may have gone to great lengths to offer this honor.

As part of the physical environment, people can limit access to food, particularly in the context of conflict. Food is thus of explicit significance to Marine operations. Control of food during times of scarcity is a central lever of influence in counterinsurgency (COIN) and other hybrid war conditions. This is because food scarcity can be real or created, in order to control people and their loyalties. For example, during 1932-33, a famine in the Ukraine killed over six million people. The Soviet government caused it, to eliminate nationalist sentiment and segments of the peasant population opposed to collectivization.

In related fashion, Marine personnel involved in provision of food in a disaster relief or COIN environment will find that local communities have expectations with respect to food availability and distribution. These expectations relate to quantity, kind of food, and who should provide it. Fulfillment of local people's expectations can determine legitimacy of outside forces. In OEF, for example, Afghan National Army soldiers have come to expect food provision from the coalition. However, Afghan soldiers do not respond well to MREs, with some considering this an indication of American condescension.[8]

Culture Operator's Questions: Food

- What are the local staples, and what is the required labor to grow, prepare, and serve them?

- What kinds of locally-accepted foods are considered strange, dangerous, or not even food to less-traveled Marines?

- What foods—eaten either by U.S. personnel or by local people—are so out of place as to raise concerns about health or sanitation?

- What foods are served by whom, to indicate the status of server or guest?

- How do Marine operations or logistics impact the ability of local people to obtain essential foodstuffs?

•   What foods have which kinds of ritual significance?

•   What are the time- or calendar-related roles of what kinds of foods?

•   Which foods are strategic commodities, inasmuch as controlling access to them influences one's coercive or political power?

•   What, in local terms, is considered food sufficiency, food scarcity, and the proper role of external forces in providing food?

Materials for Shelter. Around the world, people build their homes to fit their environment. Even in the U.S., builders prefer to use local materials that are easily obtainable and suited to the climate. Thus, people in the Pacific Northwest build homes out of pine and cedar—wood that is plentiful, affordable, and allows homes to breathe in a wet climate. In contrast, in southern California and the Southwest, where trees are hard to obtain and the climate is dry, the preference is for adobe houses which stay cool in hot climates.

Not only do cultures use local materials suited to their climate, but building styles and locations reflect experience in living in the region. In areas subject to floods for example, people have learned to put houses on stilts or to live inland away from the shore. In some desert areas, local people may build their houses with doors and windows facing away from prevailing winds to reduce the amount of dust infiltration. In constructing buildings in an area of operations, Marines should observe local building techniques to avoid unanticipated disasters, such as setting up a base in a traditional flood zone.

Likewise, particularly in civil-military operations (CMO), commanders need to choose or contract for materials available (and thus maintainable) locally—and that the local climate can sustain. It is also usually better to use materials and workers based on the entirely local economy, because especially at the company and battalion level of operations in the expeditionary and irregular environment, it is the local indigenous population who has the vote of most importance to the success of Marine activities.

Example: Materials, Sustainability, and Credibility

After the overthrow of the Taliban in Afghanistan, U.S. forces let out contracts for the reconstruction of Afghan towns and villages. A focus involved schools and clinics. The hope was that better education and health, and the positive attitude towards the U.S. they would bring, would deter a Taliban resurgence. However, the company building the schools and clinics in a particular province used materials procured from outside Afghanistan, and laborers from outside the province.

Furthermore, they selected building sites ill-attuned to the local interaction between climate and terrain. Therefore, what were built as shiny, modern classrooms and clinics, within less than a year, had sunk into soil unable to accommodate them, had been eroded by snows and rains, were too heavy to warm, and then started to disintegrate, becoming useless hulks of buildings a year later.

At the same time, their construction had not improved local incomes, and the people who were to inhabit them—doctors, nurses, and teachers—were not from the area, and were disinclined to leave their home provinces or cities.

Ultimately, the message to the illiterate though intelligent local Afghans was that the priority of the coalition in Afghanistan was not Afghans, and that coalition forces were similar to previous here-today-gone-in-a-few-years outside occupiers. Thus, ignorance of local climate, terrain, and building materials had concrete effects of hampering local education and health, and second-order effects of undermining U.S. credibility in the province.

Using locally manageable and legitimate materials will gain greater credibility not only for Marines in the AO, but for the overall operational plan; conversely, one can sink in the mud of good but misunderstood intentions if the physical environment's materials for building and shelter are not considered.

Beyond the functionality of buildings and building materials, shapes and architectural styles also communicate culturally-coded messages about proper use of buildings, and status of the occupants or owners. For example, in different cultures, certain building can or cannot be used for certain activities.

Example: Culturally-Coded Building Use

In Christianity, churches are legitimate places to mark life-cycle events: birth, baptism, marriage, and death. So, weddings, funerals, and so on, are supposed to occur in churches. Likewise, the purity of a church is not violated by the introduction of things like blood or corpses into that church. This also means that in Christianity, events that occur in churches are less likely to occur, or less accepted when they do occur, outside of churches or church-like atmospheres. An indoor basketball court, for example, is a less preferred site for a wedding, funeral, or confirmation.

By contrast, in Judaism and Islam, synagogues and mosques are places of worship and study of scripture and law. In these religious traditions, they are not locations to mark life-cycle events. In traditional Judaism and Islam, therefore, weddings, funerals, circumcisions, etc., are not to occur in mosques or synagogues, while they can occur in general-purpose buildings appropriately decorated—a cleaned up, decorated indoor basketball court, for example.

Likewise, prayer places are to be kept pure; meaning it is improper to use them as infirmaries, catacombs, or in extremis morgues—unlike in certain South East Asian ritual sites. So, in different cultures, different buildings have different functions.

Building styles and the use of architectural space, particularly in the urban environment, are important cultural aspects of the AO with significant tactical and operational implications. As just one example, house architecture, building uses, materials strength, and construction styles became essential squad-level knowledge during the combat in Fallujah in 2004. Marines developed this knowledge themselves and integrated it into battle planning and operations.[ix]

Culture Operator's Questions: Materials for Shelter

•   How do structures fit the geographic, climatic, and physical aspects of the environment?

•   What do the internal and external appearances and materials used in structures communicate about building purpose, occupant status, etc.?

- What materials for building, repair, and maintenance are local to the AO?
- What are the central tactical implications of building styles, neighborhood layout, etc?

Climate and Seasons. Climate and the changing seasons may not only affect local building styles, but also cultural lifestyles and activities. In hot climates, such as Mexico and Greece, the social pattern of the noon-day "siesta" sets a daily rhythm that keeps people out of the sun during the middle of the day. Most economic and social activities in these countries occur during the morning and evening instead.

Accustomed to our American pattern of intense work from "nine to five," Marines may find these cultural patterns frustrating—especially when hiring local people for work projects. However, if Marines can include a cultural group's daily rhythms into their operational plans, local people may be much more cooperative. Realistic expectations can be set regarding the hours local people are willing to work, and schedules and target dates are more likely to be adhered to by partners in the AO.

Like climate, seasons often have a major effect on cultural rhythms. In cultural groups that are closely tied to the land, spring and fall may bring specific activities such as planting and harvesting. For groups that herd animals, such as the Bakhtiari of Iran or the cowboys of the American west, spring may be the time of migration to greener pastures high up in the mountains. Even industrial societies have their seasonal rhythms. In Europe every government organization and business empties out in the summer as people head off on their summer vacations.

Marines need to understand these seasonal variations in order to plan operations that depend on the labor of local peoples. Otherwise, they may suddenly discover that all the able bodied men have disappeared for the harvest, just when a major Marine construction project is about to begin.

Likewise, Marines need to consider both season and climate when planning operations with indigenous forces, or when fighting against local armed groups. As just one example, U.S. personnel in Afghanistan frequently notice a drop in insurgent activity during the winter. Over time, they have come to understand that this is related less to diminished insurgent enthusiasm for anti-Afghan violence, and much more to the local cultural disinclination to fight during the winter months in high altitude. Conversely, the upsurge in violence over the summer and autumn is seasonally driven, and not necessarily a function of greater insurgent zeal.

Culture Operator's Questions: Climate and Seasons

- How does the climate influence local attitudes to—and capabilities for—work, business, and combat?

- What is the relationship between climate and season, on the one hand, and battle rhythm and operational tempo, on the other?
- What, in local terms, passes for good weather, bad weather, etc?

Fuel and Power. All societies need fuel to cook, to heat, and to provide light. As the taste for Western goods such as satellite TVs, cars and washing machines spreads to the most remote corners of the globe, obtaining power—and in particular electricity and gasoline—has become a major concern of most communities.

Today every state is expected to supply electricity and gasoline to its inhabitants, at least in urban areas. However, electric and gasoline supplies are frequently unpredictable in developing countries—and even occasionally in highly developed countries. Shortages of gasoline can leave cars stranded, and power outages are a common occurrence in many countries. People living in these conditions often view the state as incapable, thus reducing not only the state's prestige, but also its legitimacy and then authority, as well.

In addition, many third-world cities have large populations of un-official squatters that surround the city, living in conditions of extreme poverty without electricity, heat, water or sewage. Mumbai, in India, is a perfect example of this. At the same time, these "un-official urbanites" surrounding cities can tap into electricity and fuel grids to power TVs, refrigerators, and even internet cafes. This reduces the amount of power available to the city, and creates semi-legal facts on the ground that governments have difficulty changing.

As an example, the Gecekondus (squatter settlements "put up over night") surrounding Turkish cities have developed over the past generation into their own established quarters with material desires associated with legal urban areas. Some have attained legal status; overall, the building materials are poor, the stress on the power grid is great, and in times of power outage or earthquake (as in 1999), the political regime is held accountable.[10]

This means that inadequate and unequal provision of power is frequently a cause of frustration for local inhabitants who blame the power failures on the corruption and ineptness of the government (not without good reason). The opposite is also true: Availability of power, but no way to use it, can cause just as much frustration with governments. In Turkmenistan, for example, oil is extremely cheap, as it is produced locally in great amounts. Turkmenis can thus buy it for a pittance. Yet, so few have cars, and so few can open businesses able to take advantage of cheap fuel for long-distance travel, that the same effect of a shortage of fuel can be seen in economic and political terms.

When deployed and conducting civil-military or humanitarian operations, the ability (or inability) of Marines to provide predictable power may become a symbol of the U.S. military's power—or willingness—to control essential resources upon which local people depend. Recognizing the symbolic importance of power in providing legitimacy to the local government or military, insurgent groups may target power plants in order to disrupt service and discredit U.S. operations in the region.

Facing an environment of unpredictable (or in remote regions, non-existent) power, many communities rely upon traditional solutions to their fuel needs. Wood continues to be an important source of heat and fuel for people both in cities and the countryside around the world. Cultural groups that live in arid regions with little wood, such as the Masai of eastern Africa or the nomads of Mongolia, use the dried dung of their cattle or camels to ignite cooking fires. And the Inuit (Eskimos) of northern Canada as well as some remote Scandinavian communities burn dried peat to heat their homes. Since wars frequently destroy local forests and resources, Marines need to assess the effect of military actions on the ability of local peoples to gain access to essential fuel sources.

Culture Operator's Questions: Fuel and Power

•   What are the locally found, or locally produced sources of power and fuel?

•   What is the relationship between local elites and access to/provision of fuel and power?

•   How does the larger government authority provide, or control, access to power?

•   What do local people expect of outside forces in terms of power/fuel provision and protection?

•   What are local work-arounds to deal with shortages of power and fuel, and how do Marine operations impact them?

•   What local issues regarding power and fuel are overshadowed by more pressing operational considerations?

Notes

[1] See David Michaelson, "From Ethnography to Ethnology: A Study of the Conflict of Interpretations of the Southern Kwakiutl Potlatch," Unpublished Ph.D. Dissertation, New School of Social Research, 1979.
[2] See, for example, Elisha Kally and Gideon Fishelson, *Water and Peace: Water Resources and the Arab-Israeli Peace Process* (Westport, CT: Praeger Publishers, 1993); R. Hrair Dekmejian and Hovann H. Simonian, *Troubled Waters: The Geopolitics of the Caspian Region* (London: I. B. Tauris, 2003).
[3] John F. Kolars and William A. Mitchell, *The Euphrates River and the Southeast Anatolia Development Project* (Carbondale, IL: Southern Illinois University, 1991).
[4] Adeeb Khalid, *Islam after Communism: Religion and Politics in Central Asia* (Berkeley, CA: University of California Press, 2007).
[5] Neville Mandel, *The Arabs and Zionism Before World War I* (Berkeley, CA: University of California Press, 1980); Kenneth W. Stein, *The Land Question in Palestine, 1917-1939* (Chapel Hill, NC: University of North Carolina Press, 1984); Issa Khalaf, *Politics in Palestine: Arab Factionalism and Social Disintegration, 1939-1948* (Binghamton, NY: State University of New York Press, 1991).
[6] For more on local relationships to land in Africa of concern to U.S. Forces, see Major Christopher H. Varhola (U.S. Army Reserves) and Lieutenant Colonel Laura R. Varhola (U.S. Army), "Avoiding the Cookie-Cutter Approach to Culture: Lessons Learned from Operations in East Africa," *U.S. Army Military Review* (December 2006): 73-79; Christopher H. Varhola, "The Multiple Dimensions of Conflict in the Nuba Mountains of Central Sudan," *U.S. Army Military Review* (May-June 2007): 46-55.
[7] J. Wolch and M. Dear, eds., *The Power of Geography: How Territory Shapes Social Life* (Unwyn Hyman, 1989).
[8] "Debrief of BN-ETT to Afghan National Army, May 2005," Conducted by Barak Salmoni (Center for Advanced Operational Culture Learning, May 2005).
[9] Catagnus, Edison, Keeling, and Moon, "Infantry Squad Tactics," *Marine Corps Gazette*, September, 2005.
[10] Umut Duyar-Kienast, *The Formation of Gecekondu Settlements in Turkey: The Case of Ankara* (Lit Verlag [ABC], 2006).

Chapter 4

Dimension II
The Economy of a Culture

All cultural groups have a specific system for obtaining, producing, and distributing items (food, clothing, cars, houses, etc.) and services (medical care, education etc.) that people need or want to survive in their society. This system is called the **economy of a culture.**

When Americans think of the word "economy," we tend to think of money, stocks, international banking and trade, and perhaps the national debt. These words all describe aspects of one kind of economic system: a system based on the exchange of symbols (such as money, checks or credit card numbers) for goods (such as cars, food, clothing) and services (such as dental care or a college education).

Typically this system is regulated, taxed, and measured by our government. In short, this is a completely monetized, industrial economy. In fact, in its latest stage of development emphasizing high technology, intellectual capital, electronic finance, and the disregard of international boundaries in the definition of "economic interests," etc., one can consider the West as exhibiting characteristics of a "post-industrial economy."[1]

What people growing up in the industrial or post-industrial West recognize most readily can be termed "the formal economy." The formal economy is the subject of study for an entire discipline called "Economics." However, in the context of operational culture for the expeditionary warfighter functioning among many cultural groups, this approach to economy will not be the focus of the current discussion. Here, we instead examine the many other ways that economies are structured, using a cultural perspective.

As this chapter illustrates, there are numerous economic exchanges that never use money. Furthermore, much of human economic interaction is not regulated, taxed, or measured by national governments. For the Marine working in a foreign AO, understanding and working with these other forms of economic exchange may be critical to success in local operations.

There are three important models of economic systems that are significant for a Marine working in a foreign AO:

- Formal and informal economic systems
- Economy as a network of exchange
- Economy as a way of structuring social relationships

Each of these models provides a different approach to understanding economic relationships. As a result, all three models can be seen as explaining economic behavior using complementary rather than mutually exclusive perspectives. This means that in any AO:

1. There will be an informal economy
2. The informal economy will exist intertwined with the formal economy
3. Economies will work as networks of exchang
4. Economic interactions will structure social relationships

Thus no one approach that we will examine here provides a complete picture, and Marines will probably find they use all three methods of analysis when working in a new region.

Formal and Informal Economic Systems

Much has been written about the formal versus informal economy. Although a multitude of descriptions exist, for the purposes of this discussion we use the following definitions:

Formal Economy: Those economic interactions and exchanges that are regulated, taxed, tracked and measured by a state government.

Informal Economy: Those economic interactions and exchanges that are not recognized, regulated, controlled, or taxed by a state government.

The concept of an "informal economy" is based on certain assumptions, which influence the way that people view economic systems around the world. Some of these assumptions:

- There is a "formal," governmentally regulated, proper economy. "Informal" economic activities are outside of this realm— in other words, these economic activities are "not formal" and often "illegal."

- The formal economy is somehow the "real" or "legitimate" or "valid" economy, while the informal economy is a marginal, aberrant system that exists illegally on the periphery of society.

- The informal economy is the sign of a weak state that is unable to bring the economic activities of its inhabitants under its control and supervision.

- In an ideal world, a strong state would have no informal economy and be free of the many illegal and extra-legal economic activities that currently occur outside of its control.

In this chapter, we provide an alternate model of the economic system of modern states, arguing that **informal economic activities are actually a permanent and normal feature of all human behavior.** In fact, "informal" economic ways of interacting have always existed in *parallel* to—and sometimes have even been *central* to and *animated*—the formal, governmentally preferred system.

These conclusions are obvious when one realizes that formal states with the ability to control the economic activities of their populations have existed, at best, for only a few hundred years—and then only in specific locations for limited periods of time. Yet people have been living in groups, and obtaining and exchanging the food and other goods and skills they need for survival, for millions of years. On the human time scale, people have been actively maintaining non-formal economic exchange systems for millennia before states developed to try to regulate them.

Contrary to the way scholars have labeled these systems, formal economies are in many ways the exception to the rule—they arose late in human development and were superimposed upon existing "informal" systems. Today people continue to use these non-formal systems, despite the efforts of states to impose new, externally controlled systems. Why, then, would we maintain the reverse view that the formal economy is the true and optimal economy? The answer lies in looking at the needs of the state, and the answer's components are all important to a Marine.

First, from the state's perspective, unregulated activities are untaxed activities; hence a state with a large informal economy and a small formal economy will lose much of its potential tax revenue to an unregulated, untaxed system. In civil terms, the state will lose the capacity in those very areas important for its legitimacy: provision of welfare, essential services, and a society-oriented safety net.

As the successful British provision of land, building materials, schools, medical care, and subsidized local governance to (particularly ethnic Chinese) civilian Malayans during the "Malay Emergency" in the 1950s showed, these are all things that give people a stake in the state, or status quo, as opposed to other options. By contrast, the Egyptian state's inability to provide a social safety net during the Islamist insurgency of the 1990s resulted in an apathetic citizenry, so that the state had to rely on high levels of force, and a hope that the Egyptian "nature" was more opposed to violence than it was to a harsh state.

Second, in military terms, unregulated economic activities mean that the state will have less funding at its disposal to raise and sustain military forces—the state will have lost the financial component of "the people" from Clausewitz's "remarkable trinity" which we discussed in the Introduction and Chapter One of this book.

Third, and most significant for the Marine, unregulated economic activities pour money, goods, and power into non-state sectors (including but not limited to drug cartels, crime rings, and insurgent movements) which can act as a threat to state stability.

Marines working in a foreign AO will find that two important features of the informal economy may have a significant impact on operations:

- The goods and services provided in the informal economy
- The people who participate in the informal economy

Goods and Services in the Informal Economy

There are three categories of goods and services in the informal economy:

- Illegal goods and services forbidden by the state
- Quasi-legal goods and services
- Goods and services that are ignored by the state

Illegal Goods and Services. For the Marine, the informal economy is especially significant since all illegal goods and services move through this system. It is important to note here that the distinction between illegal and legal goods and services is not always easily defined. A gun sold in a sporting goods store with a license is a legal good; yet when smuggled across the border to be sold to an insurgent group it is an illegal good. Certain goods or services may be forbidden in one country but permitted in the neighboring country. Alcohol is forbidden in Saudi Arabia but permitted in Egypt. As another example, prostitution and marijuana are legal in Holland but illegal in France.

Furthermore, even within the same country, regional differences exist in the willingness of officials to prosecute people for selling certain forbidden goods or engaging in illegal services. In parts of California, for example, the possession of small quantities of marijuana has historically been ignored, although the U.S. national government forbids the drug.   Finally, some goods may be illegal and then circulate through the economy to become "legal."   Stolen goods that emerge in pawn shops and second-hand stores fall into this category.

The ambiguous nature of illegal goods and services often can lead to confusion in military operations. Marines need to be familiar with local laws and customs regarding the legality of certain goods or behaviors.

One category of illegal services provided through the informal economy is especially relevant to Marine operations: bribery. Bribery, in economic terms, is the purchase of a service (obtaining a government document, getting medical care, getting out of jail) from an official representative who does not have the right to charge for such a service. In the U.S., bribery is considered a serious crime. However in many countries, bribery is an accepted, and perhaps necessary, way of doing business.

Example: Bribery as Accepted Business as Usual

During the Communist era in the Soviet Union, many basic goods such as food and clothing, and services such as medical care and building repair, were extremely scarce.   In the Communist economic model, all workers earned the same minimal pay regardless of their skill or effort. As a result, an informal economy developed where virtually every worker earned their survival pay from their day job, and then supplemented their income through informal activities. For many, their "real income" came from the bribes they received due to their strategic position in their jobs. Storekeepers accepted bribes to inform individuals of the arrival of new goods; doctors accepted bribes to offer individuals urgently needed medical care; and government officials took bribes to help individuals locate scarce housing. Despite the end of the Communist model of economics, Russia today continues to operate on this dual system of day job supplemented by "special pay or tips for services rendered in the course of doing one's job. Bribes continue to be a necessary part of daily life.

In many developing countries, the extreme red tape and bureaucracy involved in obtaining governmentally required papers and documents is often accompanied by the practice of officials accepting "donations" to speed the process. This expensive and slow system is often a causal factor in creating the exchange of quasi-legal goods and services in the informal economy.

Example: Government, People, Illegal and Legal Economy

The interaction between legal job and informal illegal activity can be seen in modern India. The Indian constitution grants almost iron-clad job security to employees of the India Administrative Service (IAS). It is almost impossible to fire them. Therefore, Indians of all castes—even those who have been abused by the Service—would like a job in it. It is thus normal for aspirants or their parents to pay a "fee" for admission to the IAS. Once in the IAS, civil servants find they have a guaranteed paycheck, which, however, is lower than that of private-sector professionals.

IAS employees therefore have every incentive to require "fees" from Indians who come to them to receive services that are supposedly a part of normal IAS duties. The IAS is in fact chock-full of license-, permit-, quota-, and reporting requirements, and each one of these is an opportunity for a citizen to bribe an IAS member for good favor.

Additionally, IAS officials use their formal economic and legal authority not to constrain the informal economy, but to enter into it for material benefit. For example, the state has authorized 99,000 permits for bicycle rickshaws in New Delhi (for which people often have to pay bribes). However, more than 500,000 rickshaws are being operated. Rather than raising the quota to reflect reality, or confiscating the illegal rickshaws, the IAS and Police ensure that over 100,000 Indians work illegally—in return for regular bribes.

The same is true with the city's 600,000 street hawkers. It's not clear if this is a legal or illegal practice, but in spite of middle class citizens' complaints about street-hawkers, very few are arrested. Rather, they regularly pay a "fee" for occupying public space, and the police regularly raid these peddlers, "confiscating" their goods—which often make it back onto the street in sales networks that pay kickbacks to the IAS and Police.[2]

Quasi-Legal Goods and Services. These are items and labor that are not illegal in principle. However, like the rickshaws we saw above, in order to be permitted by the state, they require taxation, registration, and regulation. For example, the sale of food and clothing on the street by a licensed and taxed vendor in New York City is legal. However, since licenses are limited and may be too expensive and difficult to obtain for an individual starting without much capital, many street vendors offer their goods without licenses. In much of the developing world, a significant percentage of the labor force (up to 40-50% in some countries) works in this informal sector, providing unregulated goods and services that in theory could be legal.[3] Any Marine who has traveled outside of the Western world has observed vendors on the side of the road or in open air markets selling anything from home-made bread to used clothing to car parts. The majority of these vendors provide their goods without government regulation.

Studies suggest that a major reason vendors do not obtain licenses is the difficulty and expense of going through the governmental bureaucracy.[4] Added to this is the common expectation of paying bribes to move one's paperwork through the system in many countries, making the application for licenses almost futile to all but a few influential members of society. For example, in Lima, Peru, it took almost 300 working days—thirty-two months of income in the local minimum wage—to register a business with only two sewing machines![5]

Like the sale of small scale goods, many of the services provided in developing—and for that matter developed—countries fall under this category of quasi-legal status. Frequently these services are provided by categories of people who are marginalized from the formal economic structure. In the U.S., for example, Americans often hire new immigrants for services without asking them if they are licensed or taxed by the state. These immigrants can be found working as maids, gardeners, construction workers and agricultural laborers, among other jobs.

Goods and Services Ignored by the State. Due to the impossibility of regulation and the view by the state that certain economic activities are insignificant or not recordable, certain goods and services are deliberately ignored by the state. In the U.S., for example, teenage babysitting is generally ignored. Unpaid labor—such as seasonal agricultural work by family members or a wife's management of the books of her husband's business—also is largely unregulated in America.

The weaker the state, and the greater the distance between a community and the taxing/regulating center, the more goods and services go unregulated. One sign of a partial or complete state failure in a region is the lack of effort to tax or regulate goods and services in that area. This is not to imply that local residents get a free ride. To the contrary, often a criminal, warlord, etc., steps in to fill the taxation gap, extorting "payments" from the locals in the region in exchange for the security and protection that the state no longer provides.

People Who Participate in the Informal Economy

The informal economy is populated by people who, due to traditional social roles, legal strictures, or the prevailing ethno-religious balance of power at the formal/state level, do not dominate in the visible, monetized economic spectrum. Four categories of people are commonly found in the informal economy:

- Women
- Children
- Non-Citizens and Illegal Immigrants
- Ethnic/Religious Minorities.

Women. Women commonly work in the informal economy because many informal economic activities are compatible with their domestic duties. A woman can take in wash, bake bread or other food for the market, grow vegetables in her garden, baby-sit, and even do piecework and sewing in her home while watching her children and taking care of the house. In communities which practice the seclusion of women (as in many Muslim regions; see Chapter Five), such home-based work has the added advantage of allowing the woman to earn money while remaining out of "public view." Since these domestic activities are sporadic and not easily monitored by the nation-state, such home-based activities typically fall under the quasi-legal status of the informal economy.

Example: Women as Informal Economic Agents

Many studies of the Middle East and North Africa suggest that women have the lowest rates of employment in the world. Current estimates suggest that women form only 4-22% of the labor market in the Middle East and North Africa; this is less than half of the rate for women in Europe and the U.S. These figures, however, overlook the significant roles that women play in the informal economy in the region.

In North Africa, for example, women not only provide unrecognized labor in agriculture, but also produce handicrafts and food that are sold by males in the market (see Chapter Five); run small businesses from their home (such as sewing and hairdressing) and offer traditional services to other women (as midwives, soothsayers, teachers and traditional hennanas). Such work provides essential income to their households without challenging the man's culturally-mandated need to appear as the sole breadwinner and supporter of the household.[6]

Therefore, in a foreign environment, Marines may encounter women working at home, performing economic roles that appear exploitative, as they receive little or no pay. Other factors, however, such as attitudes, social relationships, and locally-accepted economic practices, might in fact make it better for women to work from home, in the informal economy.

For example, if Marine forces were to establish a formal factory setting for sewing, where women could obtain regulated employment and be protected by the state (or civil affairs unit), U.S. personnel might think they had prevented exploitation of women. Instead, they might discover that women do not come to the factory. This could be because working out of the home was more expensive for the woman, for many reasons. At the same time, a sewing factory that takes women out of the home would disturb local economic relations in the region based on a set of attitudes and practices.

Children. Since many countries forbid child labor, children's economic activities tend to be restricted to the informal economy. Although we frequently assume that child labor is a sign of poverty in developing countries, the reality is that many children in the U.S.

also participate in the informal economy. Running lemonade stands, mowing the neighbor's lawn, and babysitting are all economic activities in which American youth engage without paying taxes or obtaining the necessary licenses from the state.

In many developing countries, however, children's income or labor contribution is often *essential* to the survival of the household. Children help in the harvest, sell products in the market, or work in family businesses such as rug weaving. This income is counted upon for the family budget. Although we may not approve of child labor, in planning, Marines should still evaluate whether their actions could unexpectedly restrict children from engaging in economic activities that support their families, and, as we will see in the next chapter, reflect different cultural notions of age-appropriate behavior.

Illegal immigrants. Like children, illegal immigrants are also forced to work in the informal economy because the state forbids their participation in the formal economy. Paradoxically, while all nation states make laws forbidding certain individuals to work within their borders, the formal economies of virtually every country around the world depend upon the labor of illegal workers. Illegal immigrants generally undertake the work that legal residents of the country find distasteful or undesirable, whether it be ditch digging, cleaning toilets, or picking fruit. Without the participation of illegal immigrants, these jobs would not be filled—again, formal and informal economies interact, the former requiring the latter.

Because illegal immigrants are not protected by state laws, employers are able to pay these workers at low wages and ignore benefits and protections that legal citizens enjoy. Although such arrangements are clearly exploitative, the formal economy of the country gains from the lower costs of products and services that are provided from this cheap labor.

Due to their marginal nature in society, as well as their greater mobility, illegal immigrants are also a target recruit pool for illegal labor activities ranging from prostitution to drug dealing to arms smuggling. As an immigrant group grows, migrant sub-communities based on region of origin often develop in certain areas. Due to their marginal economic and social status, migrant enclaves may serve as excellent recruiting grounds for insurgents. As we will see in Chapter Six, illegal immigrant laborers are by definition left out of the political structure, and thus may seek to challenge it.

Ethnic and Religious Minorities. Like immigrants, ethnic and religious minorities may turn to the informal economy because cultural prejudices or laws make it difficult for these groups to work in the formal economy. In order to avoid special taxes or licenses due to their ethnicity or religion, members of such minorities may choose to engage in work away from the regulation or control of the formal government.

Particularly in areas where ethnic and religious groups may cluster in specific quarters or regions, a strongly independent non-regulated informal economy may develop on the lines of religion or ethnicity. Although a few larger businesses in the community will follow the state rules, the community as a whole may engage in business and economic transactions—such as the sale of imported or homemade goods, offering of traditional cultural services, or provision of illegal labor—that are predominantly unregulated by the state. Sometimes these communities become specialized in certain economic niches—for example the Sikh taxi driving communities in the U.S.—that mix formal governmental activities (registered taxis) with informal personal economic exchanges and family ties, as we discuss in the following section. Often, groups such as this are doubly driven to disadvantaged informal economic relationships, as both minorities and immigrants.

Example: Illegal Minority Immigrants in a Formal-Informal Economy

Throughout the U.S. Chinese restaurants functioning in the formal economy employ people from the informal, unregulated, semi-legal economy. In particular, the coastal province of Fujian, China, is a hub for immigrants to New York. Many of these immigrants are undocumented, and arrive to the U.S. through human smuggling networks. People who run these networks charge about $70,000 to transport an aspiring worker to the U.S.

Example: Illegal Minority Immigrants in a Formal-Informal Economy; continued.

Once here as an illegal immigrant with no English skills, a Fujianese cannot benefit from American legal, economic, or social welfare protections. Instead, they rely on an informal, semi-legal labor exchange, taking the form of listings on chalkboards in restaurants and transient dwelling areas. Negotiating directly with restaurant employers, they then rely on an informal transportation network, of cars and vans that reaches from New York to the Midwest, to arrive at restaurants for menial work, or work as chefs. These new laborers, unregulated by the state or local authorities, live in groups, often nearby or on the second floor of restaurants, working six-day weeks at twelve-to-fourteen hours a day.

This intensity of work usually burns out laborers after several months. They then return to New York, beginning the cycle again, since it will take up to ten years to pay off the debt for being smuggled to the United States.

Thus, an actor in the formal economy, in this case a Chinese restaurant in New York, Chicago, or Columbus, will have subordinate economic actors, some of whom are legal immigrants, and others of whom are illegal ethnic minority immigrants, whose entire economic and social welfare net hinges upon the informal, semi-legal economy.[7]

Culture Operator's Questions: the Informal Economy

In a foreign AO, Marines will need to ask the following questions, engaging in relationships with local people based on the answers:

- What categories of people work in the informal economy?

- On what commodities/services does the informal economy focus?

- What is the relationship between the informal economy, on the one hand, and unregulated movement of people, crime, and violence, on the other?

- How will Marine operations impact the informal economy and the people in it?

- How will the Marine impact on the informal economy influence attitudes of certain sectors of the population to the Marine presence?

- How does the formal economy rely upon the informal economy, and what abuses of the AO's population does this cause?

- What opportunities exist for the population, based on the formal economy's relationship with informal economic practices?

- What are formal/informal economic actors' expectations of the state or over-arching political-military authority, with respect to involvement in or disregard for economic activity?

- What is considered an "illegal" good or service in the AO, on what basis?

- What goods/services are legal, but culturally frowned upon? Who deals in these goods/services?

- How will Marine expenditure in the local informal economy, or employment of local informal economic actors, influence the socio-economic balance of power in the AO?

Economy as a Dynamic Network of Exchange

At the center of cultural analyses of economic systems is the concept of exchange. One of the seminal scholars on economic systems, Mauss,[7] proposed the idea that in order to survive, people are bound in an intricate web of exchange and obligation. This network of exchange or trade creates a set of social relationships (whether equal or unequal) between people. These relationships in turn form an economic system through which needed goods and services flow. Ultimately, the pattern of exchanges determines who has access to—and control and possession of—important economic resources or wealth. As we will see in our next Chapter, this distribution of wealth is related to social structure.

Economic exchange systems are generally characterized by the degree to which exchanges are reciprocal or even. In highly reciprocal or balanced exchange systems, each person in the group owns a similar amount of possessions or wealth. At the other end of the spectrum are unequal or directional exchange systems where the wealth and resources of the community are concentrated in the hands of a group of individuals.

Reciprocal or	Unequal or
Balanced	Directional

Below, we discuss three common economic exchange patterns that Marines are likely to encounter in their operations around the world:

* Egalitarian or communal distribution
* Direct reciprocal exchange
* Symbolic directional exchange systems

Egalitarian and Communal Distribution. An egalitarian or communal economy is perhaps the simplest economic system known to man. All of the group work together to create the food, clothing, housing, or other items necessary for living, and then these goods are distributed evenly to each person. This may seem like a utopian and rather unrealistic system, and yet it continues to be common around the world today, even in countries that are clearly based on a monetary system. Communism in its idealistic form (not as it was practiced) was based on the ideals of such a system. Today, the kibbutz in Israel, for example, uses an egalitarian distribution system: All members of the kibbutz work together communally and equally share the profits of their labor.[9]

Example: Communal Distribution in the Civilian World, and the Marine Corps

In her landmark book *All our Kin*, Carol Stack describes the communal distribution system of an urban black poor community in the U.S.[10] In order to survive in a world of limited resources and poverty, members of the community participated in a web of communal distribution.

Example: Communal Distribution in the Civilian World, and the Marine Corps; continued.

In this community, communal distribution meant borrowing food, clothing, and money when needed, and giving these items away to other people whenever extra was available in one family.

Stack observed a unique feature of this poor black community which characterizes all egalitarian distribution systems: No one member or household accumulated more goods or money than any other. These systems therefore functioned as a leveling mechanism. As a result, even when one household in Stack's community would suddenly receive a financial windfall (an inheritance, a significant tax refund, winning the lottery), it was impossible for the household to hold on to it. Within a few days all of the household's kin, neighbors, and friends would stop by to claim a piece of the windfall, which was considered a legitimate practice, and which resulted in a rapid egalitarian distribution of the goods throughout the community.

Marines should intuitively understand egalitarian economies, since they also practice such a system. When actively involved in field operations, all Marines work together towards a common goal and equally share their resources. They receive the same allotments of clothing and personal items, sleep on the same beds in the same tents or buildings, eat the same food, and generally have equal access to prized resources such as the Internet and movies. Marines also receive equal access to critical services such as medical and dental care, education and training, and transportation. Indeed, this unique egalitarian philosophy of the Corps extends to officers, who, unlike in other militaries, will often wait until after all their men have eaten to serve themselves and generally refuse to sleep in better conditions or quarters than their men.

In planning, Marines need to understand that people whose economic beliefs or practices are based on communal or egalitarian distribution might react negatively to both what they perceive as unequal effort in commodity development, and what appears to them to be unequal sharing of scarce resources. In some operational environments requiring uneven distribution of resources in order to meet the requirements of the most needy, exposed, or work-engaged groups of people, Marines will need to balance this requirement with local attitudes towards egalitarian distribution.

In operations where Marines do not recognize egalitarian or communal distribution, they may be distracted by apparent "corruption" of people with whom they work. "Corruption" can sometimes be explained by a communal distribution system. This is because such a system is based on the principle that if goods or money enter the group at any one point, they will be distributed in an understood manner among the members of the group.

American personnel with deep experience working with Arab and Kurdish tribes in Iraq have commented that corruption is often in fact a necessary part of tribal functions.[11] Typically, in tribal systems which tend to be based on principles of communal distribution, there is an expectation that whatever one individual receives (whether it is a large sum of money to build a school, or a set of guns for cooperating with the military, etc.) must be shared among the members of that group. As an outsider, a Marine may not see or understand the socially required distribution of goods within the tribe.

For example, when Marines provide weapons to a police chief for use solely within the unit and in specific situations, they may discover later that the weapons have not made it to all the police offers, but have wandered into the hands of civilians of a certain tribal background. The Marine's frustration is understandable. But if the communal distribution system had been factored into operational planning, the Marine may have found that the police chief was a prominent member of a local tribe, and thus had a cultural obligation to communally distribute some of the weapons—to tribal kin, more than to police officers of other tribes.[12]

Direct Reciprocal Exchange. Like communal distribution systems, economies based on reciprocal exchange are extremely simple. Two or more people or groups come together, each offering goods or services, and exchange them. The net result from the exchange is even: In most cases, both parties in the exchange walk away feeling that they have ended up with goods or services of equivalent value.

Unlike communal distribution, which requires an ongoing sharing of goods throughout the group, in principle, reciprocal exchange can be a closed-ended interaction. The parties in the exchange trade their goods or services, and the interaction is over. This, for example, is the classic case of local markets, where people bring their goods for sale and barter or trade them for items of similar value. In such a system money is not necessary, as long as the parties feel that the exchange is fair.

Most Marines who have been deployed to Third World countries are intimately familiar with economic patterns of barter and trade with the locals: cigarettes in exchange for information; medical care in exchange for the return of stolen truck parts. In fact, barter and trade are typically the logical outcome of failed economies in failed nation states. In an instable monetary system where inflation runs rampant and local currencies quickly have no value, local people will resort (or return) to more direct ways of obtaining the goods they need.

For example, in Communist Russia, goods were scarce and the government unable to provide even basic services. As a result, an informal barter and trade system sprang up, in which skilled city dwellers would head to the countryside on the weekends and trade their services for fresh food unavailable in the city: repairing the farmer's plumbing in return for eggs; or trading electronic parts for vegetables.[13]

While **closed-ended** reciprocal exchange systems are common around the world, more significant for the Marine are **open-ended systems. In these economic systems, goods or services are exchanged *over time,* creating networks of reciprocity and obligation.** Open-ended systems are based on the principle that an item or service offered today will be returned in an appropriate manner at a later date. Open-ended systems are frequently preferred by people in close relationships, since they carry an obligation that one can call upon at a later date. In the U.S., for example, friendly neighbors may be happy to give food or tools to the people down the street, knowing that when they run out of eggs or need a hoe, they can count on their neighbor to be obliged to return the favor.

In fact, the motivating principle behind many U.S. military engagements is open-ended reciprocity and obligation. We extend military aid, loan guarantees, or grants to foreign governments based upon the expectation that they will support us diplomatically and join in military alliances and coalitions. We train foreign militaries so that they may act as partners or proxies. We build bases for foreign governments so that we may use them in time of need. We engage in civil-military operations in the hope that local populace will have a positive view of the U.S. All these are examples of open-ended reciprocal relationships, and as in such relationships, the U.S. feels justifiably short-changed if the foreign partner does not reciprocate.

Symbolic Directional Exchange Systems. Because barter and trade only succeed when the parties involved have items that they each want, the system can easily fall apart: If you come to market with chickens hoping to trade for a cow, you might go home empty handed if the owner of the cow wants to buy a camel, not chickens. As a result, many cultures have developed a form of symbolic exchange—of which the most familiar is money.

In symbolic exchange all parties involved agree that a certain symbolic item, such as paper money or coins, has a value greater than its physical usefulness. For example, many cultural groups value objects—such as cowrie shells and pigs in Polynesia or special woven cloths in parts of West Africa—that seem totally unimportant to someone outside the culture. These items are used not only for their functional value, but for their symbolic value as a form of "money" that can be used in transactions.

This discussion makes an obvious fact of American life—the exchange of a symbolic item (money) for goods and services—seem very strange. But, in fact, upon meeting the white man for the first time, certain remote cultural groups such as the Yanamamo of Brazil were completely unimpressed with the silly wads of green and colored paper that the white man carried with such respect. They could not understand why anyone could kill for this strange paper called 'money' which one could not eat, wear or use to make weapons.[14]

Symbolic exchanges are based on **two critical principles. First**, the symbolic item must be viewed as legitimate in the eyes of the people. History is scattered with cases of failed governments whose currency suddenly became worthless—the Confederate dollar after the Civil War, the German Mark after World War I. The key matter in a cultural sense is that coin and paper money do not get their symbolic value and thus utility from the soundness of a system or economic plan relying on monetary exchange. Rather, in the modern era, it is the *legitimacy* of the physical currency and the economic/political system behind it that are essential to the viability of monetary symbolic exchange systems. In many places to which Marines deploy, this legitimacy has already been undermined to a degree.

Example: De-Legitimized Symbolic Exchange Systems

Marines will remember a similar instance in the aftermath of the fall of Saddam Hussein. The Ba'th-era Bank of Iraq had been printing dinars for quite some time, and they had been in circulation as a relatively valued currency. After Saddam fell, these "Saddam dinars" lost value and legitimacy in the eyes of Iraqis, even though they could have still been used as the basis of a monetary system.

In the interim, material exchange, and alternative currencies—such as the dollar, British pound, and Euro—had greater value in the eyes of Iraqis. While Marines found "Saddam Dinars" to be great souvenirs to send home, it was the dollar (or bottle of water) that got them further as a currency. However, continuing this practice too long would have resulted in undermining the new post-Saddam currency of the Iraqi state.

Second, symbolic exchange makes it possible for certain individuals to accumulate large amounts of the symbolic item (money, stocks, or cowrie shells, etc.). This permits concentration of wealth in the hands of individuals or groups. Likewise, as individuals gain wealth, their exchanges and relationships with others become unequal. Consequently, wealthier individuals have greater access to goods, control of goods, and power within the community. As a result, symbolic exchange systems are often characterized by a directionality of interactions: Goods and services tend to flow in one direction and concentrate among certain groups, rather than moving in a circulating fashion among all members of the group.

In working in a foreign AO, Marines need to look for alternative non-monetary symbolic exchange systems. In the Horn of Africa, for example, cows are considered more valuable than money in some areas. In some countries, cigarettes serve as an alternative symbolic currency. And in certain areas, guns and illegal substances such as drugs may be a more powerful currency than money. Particularly in countries where the government's financial system is unstable and failing, these alternative currencies may actually form the backbone of the economic system.

Culture Operator's Questions: Economy as a Network of Exchange

In planning an operation, Marines need to ask the following questions about the economic systems employed by local cultural groups:

- How are important physical resources (food, clothing, shelter, cars etc.) obtained by local peoples?

- How do people gain access to critical services such as medical care, transportation, or education?

- Would a specific operational plan improve or block access to critical goods and services?

- What is the degree of (in)equity in the distribution of goods and services among the population?

- Who seems to control the distribution of goods and services, and how? Would a planned operation change this distribution pattern?

- Along with or instead of money, what do local peoples rely on to obtain and exchange goods in the region?

- If money is not the primary economic system, how could the operational plan be adjusted to use the existing alternate economic systems effectively?

Economy as a Way of Structuring Social Relationships

So far in this chapter we have discussed two distinct ways of looking at economies in cultural groups around the world: by examining the formal versus informal economy, and by looking at economies as networks of exchange. There is a third approach to examining economic behavior: a structural approach which examines the ways that economies are related to the environment in which a group lives, and how these economies structure social relationships between people. Though Chapter Five will deal squarely with the social structure of a culture group and how that social structure can influence political relationships, it is important here to begin the discussion of how economic structures and relationships set the conditions for certain kinds of social structures. We will continue this discussion in Chapter Five, extending it to political structures in Chapter Six.

As Chapter Three showed, all people need certain basic physical and social items to survive, including, but not limited to: food, water, clothing, shelter, medical care and safety or protection. On the simplest level, these items derive from the physical environment in which humans live. Not surprisingly, then, the economy of a cultural group (the way that people obtain and distribute goods for survival) is intimately related to its physical environment. For this reason, our discussion of the physical environment preceded this chapter. Anthropologists have identified four main economic systems of people based on the environments in which they live: hunter-gathering,[15] pastoralism, agriculture, and industrial production. Since hunter-gatherer groups are small and generally not important to the Marine in his or her daily work, we will restrict our discussion to the other three economic forms.

Pastoralism. Pastoralism is based on the herding of animals. Certain cultural groups have long obtained food, clothing, shelter, medicines, and protection (consider the military implications of horses) from the animals they herd. Although pastoral groups are dwindling today, they continue to subsist around the world in regions that are typically not fertile or useful to other groups. Thus, we find the greatest concentration of pastoralists in mountain regions (such as Central Asia) and in steppe and semi-arid to desert regions (such as the Horn of Africa, the lowland Middle East, etc.).

Because of the low fertility of the lands in which pastoral groups graze their animals, virtually all pastoralists are nomadic—meaning they move around and follow their herds. Frequently, pastoral groups have seasonal nomadic patterns, something that Marines in an AO can easily observe and for which they can prepare.

Adventurers in the Middle East, such as T.E. Lawrence, Gertrude Bell, and John Baggot Glubb, may have portrayed pastoralists such as the Bedouin as the "heart of the nation" based on quaint Victorian imaginations.[16] In contrast, pastoral groups are typically viewed by formal state societies as military and security threats. Because of their mobility, pastoral groups tend to ignore state boundaries; their movements are difficult to regulate; and their economic activities are all but impossible to tax. The informal economy is thus ideally suited to them.

Frequently, pastoral groups use their mobility to engage in activities not regulated by the state: smuggling legal and illegal items, including drugs, guns, and people across state borders. The Algerian War challenged France so because Bedouin and Berbers in Tunisia and Morocco could smuggle fugitives out of Algeria and weapons into Algeria, in line with age-old economic practices.[17] Likewise, the smuggling of funds, vehicles, and "foreign fighters" into Iraq from Syria, Jordan and Saudi Arabia results from organized militant groups taking advantage of economic practices of the nomadic tribal networks that straddle the borders. In this case, the pastoralists are engaging in "normal" economic activities or non-ideological criminal activity, which is then used by terror groups to their own advantage.

Pastoral peoples are often rugged fighters and scouts as a result of their years of camping and traveling in harsh physical conditions through a range of environments. Many pastoral groups, such as the Masai of East Africa and the Mongols of Central Asia, are famed throughout the world for their warrior cultures and fierceness in battle.

This has two implications. First, though we will deal with matters of identity and belief in Chapter Seven, it is significant that pastoralists cleave to these images of themselves and their character, even after they have begun to sedentarize or urbanize. In fact, Marines will most likely not find a nomadic vs. sedentary divide; rather, they will find former pastoralists somewhere along a spectrum of partial sedentarization, perhaps dwelling in a solid structure for part of the year and a tent-like one nearby during other parts of the year; or perhaps taking advantage of all the modern technological tools, such as computers, cell-phones, pick-up trucks, etc., in order to continue a form of nomadism.

Second, traits of self-reliance, independence, and the élan of "knowing the land" can be quite attractive to the governments and militaries of settled states. But it is clearly a love-hate relationship. Throughout its history the Jewish state of Israel has enlisted certain Bedouin communities into the Israel Defense Forces as scouts, sappers, and border police, counting on the "Bedouin sixth sense." However, in the Negev and Sinai regions, this same "sixth sense" has been used by Bedouin in both Israel and Egypt to collaborate with criminal gangs and terrorists in the sale and smuggling of weapons and explosives, eluding Egyptian authorities in particular. In recent cases of terrorist attacks in Sharm el-Sheikh, for example, Egyptian terrorists were supported and transported by local Bedouin communities at odds with the state.[18]

As we will discuss in the following chapters, the economic system of a cultural group is intimately related to its social and political structure. Because there is a limit to the number of people and animals that the environment can support, most nomadic pastoral groups number between 50-150 people. The small size of such a group, the difficulty of transporting large numbers of goods, and the dependence of the members upon each other for survival means that the economic relationships between the members must be fairly egalitarian. Although some members of the group may own a few more animals or possessions, in general all of the members of the group are concerned about the welfare of the members, and needy families will receive support from the group.

Typically, most pastoralists organize themselves along kinship and tribal lines—social structures we will examine in the next chapter. Leadership, authority, wealth, and status in the group are based on one's inherited position within the tribe. Due to reliance upon the group for survival and movement over large distances, many pastoral groups do not accept the concept of belonging to a country or place—i.e., citizenship as we think of it—nor do they recognize the sovereignty of state governments that try to control or subdue them. Instead, most pastoralists view their primary affiliation to their tribal group. Even today, despite the efforts of many states to settle nomadic groups, the values of an egalitarian group that governs itself and makes its own decisions persist among former nomadic groups. The many independent tribal groups in Afghanistan and Iraq today reflect this mentality of fierce independence and allegiance to their tribe, rather than to some amorphous state.

Agriculturalism. Agriculturalists produce their food, clothing, shelter and materials for exchange from the crops that they grow. In contrast to nomadic pastoralists, agricultural societies have an intimate relationship with the land they farm and are sedentary (settled) people. In an agricultural society, land becomes an important physical and symbolic commodity that is essential for survival. Conflicts over the use and inter-generational transmission of land are a logical outcome of this economic system.

Unlike pastoralists, sedentary agricultural communities are able to accumulate and store physical possessions. Agricultural societies therefore develop more complex ways of distributing goods, including inheritance of wealth and land along kinship lines. Frequently, in agricultural societies certain individuals and kinship groups accumulate a larger share of the community's wealth. This leads to a social structure in which individuals and families become stratified according to access to and control of goods, especially land. As we will discuss in Chapter Six and again for the Philippines case in Chapter Eight, this unequal distribution has important implications for the distribution of power and the resulting political structures of a society.

Another significant feature of agricultural societies is their pattern of the weekly market. In many parts of the world, people bring the food they have grown to sell in local outdoor markets. Vendors from the area come, set up a stall in an open plaza or marketplace, and spend the day selling produce or other wares to locals. Commonly, these markets have a predictable time-based pattern with different neighboring villages holding their markets on different days.

For example, on the island of Djerba in Tunisia, the market in Houmt Souk is held on Thursday; it is held on Saturdays in al-May which is 20 kilometers away; and on Sundays in the Berber village of Ajim, another 30 kilometers to the south. Similar to a trade show in the U.S., the same vendor will move from village to village, neighborhood to neighborhood, as the market itself moves. Though it may be tempting to consider these "semi-nomadic agriculturalists," they are often sedentarized people.

While many vendors may be "legitimate," in the sense of being legally licensed by the state to occupy certain stalls, often an informal economy of vendors exists on the periphery of such markets. These vendors may simply not possess the state approved licenses, or they may be selling stolen or otherwise suspect goods.

Weekly markets are an important place for socializing, exchanging information and goods, and creating and maintaining social networks. In many instances, these non-economic functions appear just as important to the market-goers as do purchasing and selling (think of shopping malls in the U.S.). As such, markets can assist Marines to create relationships with local vendors, who provide a source of local gossip (or rumor intel; "rumint"), in addition to knowledge about events and attitudes in a region. Of course, since vendors undertake a circuit from one market to the next over the period of a week or two, markets are also an excellent place for Marines to observe the flow of goods and services in a region.

Therefore, if a unit operating in the area were to impose a curfew or movement restrictions in order to interdict the flow of illegal goods through the market, this measure would also interdict the livelihoods of large numbers of producers, vendors, and consumers; at the same time it would shut off a source of rapport-building and information for Marines.

The weekly market is just one of many calendar-dependent social and economic events in agricultural societies. Because of the nature of planting and harvesting, agricultural societies have important observable rhythms. One of the ordering factors of people's lives is time, especially in agricultural societies. Frequently, religious and belief systems emerging from agricultural economies and societies incorporate time, the seasons, and the agricultural cycle itself in their rhythms and rituals. Weekly, seasonally, and yearly, people establish or are socialized to a "time-clock of life" which appears natural in that community.[19]

For example, in traditional Jewish communities, the entire day is organized as to prayer times; the entire week is aligned to the Friday-Saturday evening Sabbath; and the entire year's holiday calendar is regulated according to ancient Near Eastern harvest seasons. In Jewish communities that maintain an economic and ritual calendar tied to those planting and harvesting schedules, therefore, doing business during the fall harvest season is not possible, just as there are interruptions in the spring as well.

This means that a culture group whose environment supports agriculture will become sedentarized, as did the Ancient Hebrews as they abandoned nomadism. Settled agriculturalists are also dependent on the calendar for planting, harvesting, and marketing. Social norms and religious practices will therefore also gradually align with a sedentarized agricultural economy, and develop a closer relationship to the calendar. This is the story of Pharaonic Egypt and Ancient Mesopotamia.

Industrial Production. In both agricultural and pastoral societies, there is an intimate relationship between people and their physical environment. Although there may be some specialization (someone is a weaver, someone else is a carpenter), most people still produce at least some of the food they eat, the clothing they wear, and the buildings in which they live. To a large degree, cultural groups with these economies are fairly self-sufficient, producing and consuming what they need to survive.

In contrast to agricultural and pastoral economies, in industrial production each member of society specializes in a certain skill or task. Rather than creating everything he or she needs to survive, each person contributes a small step in a much larger economic process. By pooling their many skills, people can collectively produce a much greater variety of goods—far more than necessary for day to day survival. As a result, industrial economies create enormous surpluses.

Competition over control, movement, and sale of these enormous surpluses results in industrial economies tending to produce highly stratified social systems. Not only do industrial economies have the capacity to generate great wealth, but the social systems they create exhibit great *differences* in the *distribution* of wealth. This wealth may be distributed on the basis of one's skills, education, or family connections. As we will discuss in the following chapters, these disparities in wealth are also often connected to class, ethnic, and even religious differences.

Industrial economies, just like pastoralism and agriculture, are ultimately dependent on the physical environment. Rather than access to land or water, however, industrial production today is highly dependent on the availability of natural resources, especially energy. Likewise, many contemporary disputes between and within states center around obtaining or controlling essential natural resources. For example, though communist south Yemen and monarchist north Yemen had reunited in the late 1980s, a new civil war broke out in the mid 1990s when oil was discovered in one portion of the country, igniting suspicions of future economic exclusion in other regions. This continuing instability in Yemen provided a welcome environment to extremist elements.[20] All areas of resource dispute are likely future regions of Marine deployment, not only because of the instability they cause in key choke-point regions, but also because America's access to oil is an ever-present strategic concern.

Culture Operator's Questions: Economy as a Way of Structuring Social Relationships

In planning and operating, Marines should consider the following issues about local economic systems:

- What are the main economic systems in place in the region (pastoralism, agriculture, industrial production—all three may be present simultaneously)?

- What are the economic rhythms of the community (migration seasons, planting and harvesting, market day, work hours)?

- What are the important features of the environment that determine the economy of the AO?

- How is wealth distributed? Does wealth seem to be concentrated in the hands of certain individuals or groups? On what basis?

- How do local economic structures reflect the relationship of the group to the larger political and state system?

Notes

[1] See Charles Hecksher, et al, *Agents of Change: Crossing the Post-Industrial Divide* (London: Oxford University Press, 2003).
[2] Edward Luce, *In Spite of the Gods: the Strange Rise of Modern India* (New York: Doubleday, 2007), 78-9.
[3] International Labor Organization, "STAT Working Paper No.1–2002: International Labor Organization Compendium of Official Statistics on Employment in the Informal Sector,"
http://www.ilo.org/public/english/bureau/stat/download/compla.pdf.

[4] Ralf Hussmans, "Measuring the Informal Economy: From the Employment in the Informal Sector to Informal Employment," *International Labor Organization: Policy Integration Department* (2005).

[5] M. Estellie Smith, "The Informal Economy," in Stuart Plattner, ed., *Economic Anthropology* (Stanford, CA: Stanford University Press, 1989), 309.

[6] Paula Holmes-Eber, "Gender in the Informal Economy," *Encyclopedia of Women in Islamic Cultures Vol. 4* (Leiden: Brill NV, 2003).

[7] Margot Adler, "Chinese Restaurant Workers Protest Lost Wages," (Audio Recording) *Morning Edition*, National Public Radio, May 8, 2007: http://www.npr.org/templates/story/story.php?storyId=10089965.

[8] Marcel Mauss, *The Gift: Forms and Functions of Exchange in Archaic Societies* (Glencoe: The Free Press, 1954).

[9] Julia Chaitin, *Inside-Out: Personal and Collective Life in Israel and the Kibbutz* (Lanham, MD: University Press of America, 2007).

[10] Carol Stack, *All Our Kin: Strategies For Survival in a Black Community* (New York: Harper and Row, 1974).

[11] William Macallister, "Models of Tribal Engagement," Presentation at Center for Advanced Operational Culture Learning, 20 June 2007.

[12] "Debrief of Director, ISF Cell of 1st Marine Division," TCMEF, January 2005.

[13] Author's personal communication with Russian city dwellers, St Petersburg, Russia, October 2003.

[14] Napoleon A. Chagnon, *The Yanomamo* (Belmont, CA: Wadsworth Publishing, 1996).

[15] Hunter-gatherer groups gain their food and other essential items by hunting animals and gathering fruits, nuts, berries and roots.

[16] See Rory Stewart, "The Queen of the Quagmire," *The New York Review of Books* 54:16 (October 2007).

[17] Alistair Horne, *A Savage War of Peace: Algeria 1954-1962* (New York: NYRB Classics, 2006); M. Alexander, M. Evans, J.F.V. Keiger, eds., *The Algerian War and the French Army, 1954-62: Experiences, Images, and Testimonies* (New York: Palgrave Macmillan, 2002).

[18] See Khairi Abaza, "Sharm al-Shaykh Bombings: The Egyptian Context," *The Washington Institute for Near East Policy*, August 12, 2005: http://www.washingtoninstitute.org/templateC05.php?CID=2355.

[19] Eviatar Zerubavel, *Hidden Rhythms: Schedules and Calendars in Social Life* (Berkeley, CA: University of California Press, 1985).

[20] Stephen C. Caton, *Yemen Chronicle: An Anthropology of War and Mediation* (New York: Hill and Wang, 2005); Robert D. Burrowes, "Oil Strike and Leadership Struggle in South Yemen: 1986 and Beyond," *The Middle East Journal* 43:3 (Summer 1989).

Chapter 5

Dimension III
Social Structure

Every culture group organizes the relationships among people. The *way of organizing relationships* then defines the *kinds of interactions* people can have with each other. The resulting *pattern* of relationships can be described in terms of a *structure*. This structure places boundaries on people's behaviors, and limits access to certain people. This structure also connects individuals and groups, and defines the kinds of interactions they can have. What we have described here is a **social structure of a culture group.**

Although social structures are not immediately visible to an outsider, one can begin to grasp this concept in any environment by using the analogy of a physical structure, such as a building. In the structure of a building one finds walls—these are like social boundaries. Physical structures also possess corridors—these are like channels of social access. In both physical and social structures, these features separate people from one another or allow specific access.

Buildings are also composed of rooms which define what kinds of activities occur among what kinds of people; similarly, social structures define the kinds and nature of activities among groups of people. Likewise, physical structures often have different levels or floors; in the same way, social structures separate people into varying levels or hierarchies of position and status.

For the purposes of this book, we need to define "position" and "status."

Position: the symbolic place one holds relative to others in a social structure.

Status: the meaning and value accorded by members of a social structure to a particular person occupying a specific position in that social structure. That meaning and value derives from social attitudes to the position itself; from specific individual qualities of the person; or from both.

The above discussion of social structure, using the building motif, might suggest that structures are static and that humans' social structures do not change. If Marines were to think this, they would be mistaken and have a false impression of culture groups as being socially rigid and inflexible.[1] Just as buildings frequently undergo remodeling (on purpose or by accident) where rooms are added, combined, or removed, and new spaces created, so too, social structures constantly evolve and change whether through deliberate or accidental processes.

This tendency of a social structure to change will in fact accelerate in areas exposed to sustained and extreme political, military, and economic breakdown—as we saw from our discussion in Chapter Two of Iraq in the 1920s, 1980s, 1990s, 2003, and 2006. Since Marines tend to deploy to these kinds of places, they should not assume that an AO they have seen earlier in their career will exhibit the same social structure characteristics over time—even one year later!

Why are social structures difficult for an outsider to grasp immediately? Typically, understanding, belonging to, and functioning within a social structure occurs through socialization, often as a child. By the time a child reaches adulthood, he or she understands how relationships work in that society, and is able to navigate appropriately within various social structures. Returning to the analogy of a building, socializing a child is like conducting that child through the rooms of the house and letting him/her observe the paths through rooms and the activities in them—as well as the rooms that are off-limits. In this process, a youth is in effect *built into* the house—the child learns the social structures and then becomes part of them.

However, rules and structures that a child learns as he/she becomes an adult are generally unclear to a foreigner. In effect, the foreigner did not grow up in the same building as did the indigenous person—he is not in the same way a part of the structure of the house, and does not understand how to move through that house as well. Finishing the analogy, the foreigner is not a member of the social structure, and he does not know how to move through the social structure of that group.

Inter-cultural miscommunication is frequently the result of the inability of people from different cultures to understand each others' implicit social structures and behave accordingly. Therefore, for Marines, learning to recognize the social structures of a culture group is critical to working effectively within local political, military, economic, and social systems. This chapter will examine a number of the basic components of a social structure that Marines will encounter while deployed. It will illustrate that understanding social structures also enables one to explain axes of current as well as potential conflict among groups of people in a region.

Understanding Social Structures

Around the world, there are many different types of social structures that organize the way people interact with each other. Some of the more common social structures include: kinship (or familial) structures; business and economic organizational structures; political structures; age-grade structures; religious structures; legal structures; military structures; class structures; and social networks. Though it is tempting to think of each category (religion, economy, politics, kin, etc.) as a conceptually discrete aspect of "culture"—and though other chapters of this book address them from different perspectives—all of these categories are in fact primarily social structures.

All social structures have certain features in common:

- Social structures consist of a set of organized relationships or ties among people. In structural diagrams, these ties are often drawn as lines.

• Individuals within the structure occupy a position or specific place in relation to other people. This position is usually associated with certain activities or work tasks, and is often described by a title, such as "president," "secretary," "priest," "mother," or "third grader"—note however, that we are not speaking here of the position's political, religious, economic, or personal aspects. We focus here on the location that the position occupies in the social hierarchy. These positions are usually represented in structural diagrams by boxes or circles and are connected by the ties described above.

• Each position in the social structure is associated with a specific role. As in a play, a social role requires a person to act in certain socially appropriate ways. In the U.S. for example, General Officers are expected to be authoritative and make decisions; while fourth graders are expected to sit in their seats and fidget occasionally as they listen to the teacher. A General Officer that behaved like a fourth grader might lose his position and title; a fourth grader that acted like a General Officer would probably cause his teacher to discuss the strange behavior with his parents and send him to counseling.

• Many, although not all, social structures are organized in a hierarchical fashion: Some people occupy positions of higher status and power, while others hold lower positions of deference or obedience to those above them.

• Social structures exist independently of the specific individuals within them. People may come and go but the structures, positions, and roles continue on as new people replace those who leave. This is because a society's functionality over time depends on the endurance of relatively stable social structures—that is why conflict, war, and crisis result in destabilized social structures, which then aggravate those very conflicts and crises.

• Social structures and the positions in them are thus more important to society than are the individuals themselves who occupy them. For example, though personality and command style do make a difference, it is still the case that the relative social importance of, and behavior towards, a Commanding General remain constant over time, regardless of the occupant of the billet and his/her personality. If this were not true, the unit's structural integrity and capability would diminish.

As this last point implies, a familiar social structure for many Marines is the military social structure. Typically, this is represented by a Military Organizational Chart. This chart symbolizes—on paper—the structure of the relationships between the people in charge of a unit and their subordinates. This social structure is not a *physical* reality. Yet, the organizational chart represents a *social* reality that is very clear to the Marines in a unit, and dictates the way that people within that unit interact. People in a military unit generally know who is formally in charge (we discuss the informal lines of power in the following chapter), who reports to whom, and what work they should be performing in the position they hold within the structure of the organization. They also know what is acceptable behavior for their position and rank, and understand the consequences for their position within the unit of appropriate and inappropriate conduct. This is because a *military* unit with command relationships and staff sections is also a *social* structure with clear positions, relationships, roles, and accepted modes of interaction.

Although it is probable that most Marines have seen an organizational chart of their unit, such a diagram is not necessary or required for the Marine to know how to behave. In fact, all Marines are socialized, starting with boot camp or Officer Candidates School, to learn the rules of Marine social structure and organization. In the early months of Marine socialization they learn to recognize rank and are taught the appropriate way to respond to officers, NCOs and others above them—they are being built into the house that is the Marine Corps.

Likewise, people in different societies are socialized to learn the appropriate behaviors for their position in society and to recognize the differences in status and rank of others in their social structure. Fathers and mothers, for example, do not need kinship diagrams to understand their status in relation to their children: they understand their role from years of living in a family.

At this point it bears emphasizing that social structures are not physical things one can touch. However, they do have a direct relationship to physical structures that humans make. For example, houses around the world are laid out in ways that reflect the culturally-coded structure of the family living in it—American houses tend to be single family homes with individual rooms for family members (when income permits), whereas Asian houses are built for extended families with many people sharing one room. The former are built for nuclear family structures, while the latter reflect a social structure of inter-generational cooperation and co-location.[2]

As another example, business and military organizations arrange office size, space, and location according to the differential status and rank of the occupants, in order to dictate behavior—whether it be controlling access, granting privacy, or providing amenities to people based upon their social importance and functional roles. The **physical structure is thus built to support social roles and the overall social structure.**

The second part to this is that **physical structures influence social structures.** They can do this in at least **three ways. First**, physical space can perpetuate over time the relationships that emerge from the social structure itself. We can call this the physical environment's role in social reproduction.

Second, one can intentionally build a physical structure to create a social hierarchy and certain modes of interaction—one can use a physical space to create a social structure. **Third**, in the absence of intent, ad hoc or poorly designed physical structures can impede or change the social relations and power balance among people.

This last point is important, because if a Marine builds a *physical* structure for people in a foreign AO according to Western ideas of how people interact in *social* structures, they could 1) inhibit local interactions, thus preventing efficient work; 2) change the local pattern of interactions and thus social structure (not always a bad thing as long as it is done with intentionality and according to a mission-aligned purpose); or 3) discover that local people reject either the structure, or the work meant to be done in it—or both—thus reducing the legitimacy of the Marine presence.

Just as it may take months or even years for Marines to fully understand the rules, behaviors, and hierarchy of rank in the Corps, they must realize that understanding local social structures and appropriate interactions in a foreign AO requires time and patience. People are often not consciously aware of the rules and structures of the groups to which they belong, and they may have difficulty explaining this to an outsider. By way of analogy, knowledge of social structures is like understanding grammar. People who grow up speaking a language can use the grammar effectively, but are often unable to explain it systematically—either to themselves or to foreigners. This level and fidelity of explanation often requires someone who has learned the language as an outsider and has had to explicitly learn grammar.

Furthermore, as a member of the foreign culture's social structure, the native interlocutor is enmeshed in all the local conflicts, feuds, and assumptions about themselves and outsiders. This means that the "local guy" in an AO is often not the best authority on his own culture. He may implicitly know the rules, and know how to act appropriately, but he may not be able to make them explicit to a foreigner—he is too "in it" to step outside for the purposes of analysis, and he is too "of it" to be unbiased in explanations of who occupies what social positions in the AO. Local "cultural advisers" have too many explicit agendas and unconscious biases to be objective, and it is important for a Marine to inoculate him/herself against a "cultural Stockholm Syndrome" while deployed.

From a Marine perspective, it is worth considering that sometimes, outsiders who have systematically studied a foreign culture are more able to explain its social structures to foreigners. Again, to use the language analogy, that outsider to the foreign culture group speaks the language of his own group, so that he knows how to translate the new language he has learned to those who speak his native language.

Despite these challenges, there are a number of factors that are commonly used to organize social relationships into specific social structures around the world. This chapter will examine the central social factors that typically affect the status and position of people within social structures.

Factors Affecting Position within the Social Structure

Social structures not only determine the relationships and roles of individuals within a group, but may also organize groups into specific positions within their society. Class and caste structures, for example, generally organize entire groups of people (lower class, middle class, upper class) into specific economic and social relationships with other groups. The organization of certain social/cultural groups (such as ethnic, religious, or tribal groups) into positions of relative power over other groups has very important implications in understanding conflict and war.

Although in the U.S. people (ideally) earn their status or position on the basis of their skills and abilities, in many parts of the world this is not necessarily the case. In fact, in many culture groups, a person's roles, status and power are frequently determined by his or her biological characteristics (gender, age, race or skin color) or membership in a social group (such as tribal membership, ethnic identity, religious affiliation, or social class). As a result, a person's options in life may be dependent, not on his individual success or achievement, but on the relative status of his social or biological group. In such a situation, the most effective option for moving up in the social hierarchy may be to challenge the existing social hierarchy and/or to advocate for the better position of one's entire social group. Often the result is conflict—among classes, ethnic groups, religious groups, etc.—as one group challenges the existing system and tries to obtain greater status and power within the society. We will return to this in Chapter Six.

In order to operate effectively in a region, Marines need to be able to understand the social structure of their local communities. By understanding the roles, positions, and status of various groups and key individuals, Marines can determine which group(s) are in power due to their place in the social structure, and can also locate those individuals and groups who are in a position to influence power holders. Likewise, it becomes possible to make an estimate of those individuals and groups in the battlespace who might challenge existing power structures through either overt or covert measures, because of their disadvantaged place in a social structure that is too rigid to permit mobility.

Although there are many possible ways to categorize people, most societies around the world differentiate among people on the basis of the following characteristics:

- Age
- Gender
- Kinship and Tribal Membership
- Class
- Ethnic Membership
- Religious Membership

In the previous chapter, we discussed how these characteristics (gender, age, ethnicity, and so on) can affect the economic roles and opportunities of people. In this chapter, we extend the concept to examine how a person's social characteristics determine not only his or her economic position, but also their position in the social structure as a whole. As the reader is probably beginning to realize, economy and social structure are intimately linked. In the following chapter, we will extend the discussion to look at how one's position in the social structure then determines one's power and position within the political structure.

Thus, a culture's relationship to the environment affects its economic structure; the economic structure is linked to and reinforces the social structure; and the social structure, in turn, determines the political structure. Finally, as we will discuss in Chapter Seven, belief systems reinforce and perpetuate economic, social, and political relationships.

Age.   Virtually every culture group around the world assigns different roles, status and tasks to people on the basis of their age.   Although it would seem obvious that small children are not expected to undertake the same roles as adults, not all cultural groups define age, and age-appropriate roles, in the same way.   In the U.S. and Europe for example, people are considered children until the age of eighteen or even twenty-one. Generally, until a child reaches the age of sixteen they are also required to go to school, and not allowed to work except under special circumstances. Until children become legal adults, their parents have the right and responsibility to make decisions for their children and to support them.   In contrast, in many parts of the world, children are sent out to earn money as early as age six or seven.   In some cultures, girls may be married by the age of twelve or thirteen.

Example: Age, Labor, and Schooling

In many societies, youth are taken out of school, or perhaps never sent to school, so that they can contribute important labor to the family.   This is often the case in societies where either the subsistence- or export-oriented economy of the masses is labor intensive, seasonally or year round.

In Egypt since the eighteenth century, subsistence agriculture at the village level required the labor contribution of all able-bodied members of the family, particularly during sowing and harvesting seasons.   Twentieth-century governmental initiatives to establish schools and compulsory attendance in the rural periphery thus ran into opposition from two social groups.   The first of these were the provincial grandees who had acquired great tracts of land from the 1840s, and who used local families' labor to produce crops—often cotton—for export.   In addition to this sometimes absentee landholding class, the second group to react poorly to schools and compulsory schooling were the agriculturalists themselves.   Indebted to the landowners for seed and tools, families needed every available hand during several months out of the year.   Student registrations remained low through to the 1950s, and attrition from schools remained the rule beyond then.

Capital city administrators complained about the "ignorant peasant" who wanted to keep his children as unschooled, illiterate, and superstitious as his father and grandfather had been.   By contrast, the small number of well educated school planners of rural origins chose to focus the Education Ministry on the damaging influence of rural landowners who sought to prevent schooling in the provinces, as it would undermine their agricultural

---

Example: Age, Labor, and Schooling; continued.

profits. It was only when these younger educational planners began to adapt the schedule, calendar, and curriculum of rural schools to the patterns of the agricultural laboring class that interest in schools began to rise, along with a more positive value put on education.[3]

To extrapolate into a modern context, a schooling initiative resulting in no one coming to schools built by Marines in Kenya or Djibouti as part of stability operations would be judged a failure by Americans, and it might be tempting to view the local people as anti-American. In fact, what would be going on was that U.S. forces undertook an initiative directed at a certain age-based stratum in the culture, for whom schooling was viewed as less essential by higher status (older-in-age) members of that culture group.

Not only do children often work at a young age, but they may also participate in military and insurgent activities at ages that most Americans would consider completely unacceptable. In many culture groups around the world being a warrior is associated with manhood or adult status; boys as young as age eight or ten may be encouraged to assist in military activities—in some cases, youth at this age may indeed do so voluntarily as a way to assert manhood, or as a result of adolescent peer pressure to show manliness. These activities may range from running errands for soldiers in the field to using their small size to crawl into holes and set explosives, or carrying weapons and fighting alongside soldiers three to four times their age.[4]

Along with paying attention to the roles of children, Marines must also evaluate the cultural roles of adults and the elderly. In contrast to many Western cultures that value youth and newness as possessing the potential for innovation and progress, many societies revere the elderly. This is particularly true in Asian countries. In a culture group where the elderly are respected for their wisdom and experience, Marines have discovered that initially, local elders are unwilling to negotiate with young captains or even majors who do not have enough gray hair to command respect. In such cases, U.S. forces have at times seen benefit in including an older (and preferably grey haired) member on their negotiating team. Even when the older member was not central to the negotiations, or was a senior SNCO or WO as opposed to commissioned officer, his presence has been felt to assure local elders that the Marines take the matter seriously.

An important social structure based on age is what anthropologists call "age grades" or "age cohorts." Age-grade or age-cohort structures exist in many societies, including in the U.S. To begin to understand this, consider how Marines who went through Officer Candidates School or The Basic School together often view each other as members of the same grouping or cohort. These bonds can often be useful in the informal ways Marine officers facilitate formal interactions, assistance, or exchanges of information.

"Age cohorts," however, go beyond that. These structures group all people who are born within a certain time period (commonly one to five years). This group then moves through the life cycle together as a social unit, gaining status, prestige, and new roles as they graduate from one stage to the next. In the U.S. we use age-grading in our school systems to move children from infancy to adulthood, by sending them to a "new grade" with new tasks, roles, and expectations every year. Once the group graduates from high school or college, it is admitted to adulthood and granted new privileges, including the right to vote, to marry, and to own a credit card.

Example: Age Cohort, Identity, and Entitlement

In many African cultural groups in a belt running through the Sahel from Senegal to Ethiopia, men or women born within a certain number of seasons move as an age cohort through the life cycle. Age is not counted by a birth certificate but by the general time period (so many moons or seasons) when a certain set of children were born. In some regions in East Africa, the age group is so important that each group is given a specific name that all members bear through their lives. The group then grows up together; it goes through manhood or womanhood initiation ceremonies together; becomes eligible for marriage and parenting at the same time; and eventually moves into leadership positions in their village or tribe as a council together. Such age groups play a major role in African society, both in terms of power in the community and through networking. First, because all of the individuals in an age-cohort grow up together, bonding through various life-experiences, their childhood connections remain strong, even if they move as adults to other regions or urban areas.

Example: Age Cohort, Identity, and Entitlement; continued.

These connections then serve as important networks of power, providing access to goods, information, and people.

Secondly, each age group knows their position relative to each other in the social structure. A member of a younger age group will be expected to show deference to someone from an older age-cohort. The older age-cohorts work together as respected community leaders, making decisions for the village. Even if a person from a younger age-cohort is smart and energetic, it is understood that he will not be permitted into the closed inner circle of the older leading group.[5]

Today these age-cohort structures are beginning to break down in urban areas. Even so, Marines need to be aware that decision making still tends to be concentrated in specific age groups. In some areas, such as Chad, in fact, local people have a term for senior decision making males, whom they term "gray backs"—referring to the silver stripe on the back of a senior gorilla.

In order to plan a successful operation, Marines need to find out local cultural attitudes regarding acceptable roles for different age groups.

Culture Operator's Questions: Age

- At what age is someone considered a child or adult?

- What specific ceremonies mark the transition to adulthood?

- Which new social privileges are granted to men and women when they pass these manhood or womanhood rituals?

- What are locally accepted or expected economic roles for what U.S. society considers children?

- How should Marines prepare to respond to children that act as soldiers in militaries or insurgencies, or participate in violent activities against U.S. forces?

- What special status or roles are accorded to the elderly?

- Is there an age grading system that stratifies people according to their age and stage in the life cycle? And if so, what rights, roles, and duties do people have at each stage?

Gender. As we saw in Chapter Five with respect to informal economies in particular, with very few exceptions, every culture assigns different roles and work to men and women. Although this distinction would seem to be a natural biological one, the roles that men and women are assigned around the world vary vastly. In Europe and the United States, for example, it is assumed that women are naturally less strong than men, so that physically demanding jobs (construction, moving heavy objects, etc.) tend to be done by men. In contrast, women in a number of sub-Saharan African culture groups are believed to have much better balance and stronger heads than men. Thus, it is considered natural that women, rather than men, should carry extremely heavy loads for miles in baskets on their heads.

Labor in many societies is divided by gender, with some jobs viewed as "male work" while others are considered almost exclusively "female work." In some culture groups, men may refuse certain tasks such as cooking, washing clothes, or taking care of children because they are considered female work and demeaning. In other regions, taboos and beliefs about women (especially women who are menstruating or pregnant) may mean that women are forbidden to work with or touch certain objects such as knives or brooms or to engage in activities such as heavy lifting.

These beliefs and attitudes may affect the kinds of work local populations are willing to undertake for Marines. They may also affect the attitudes of local people towards the work that male and female Marines undertake. In cultures that view cooking as a female task, for example, a male Marine cook may be viewed as effeminate by local populations.

Attitudes about the appropriate work for men and women may not only affect Marines' interactions with foreign local populations, but also with their militaries. Although the U.S. has recently begun to include women in the military—and then not in the infantry—many other culture groups hold a long and ancient tradition of women as warriors, including as weapons-bearing combatants. The list of female combatants and military leaders is not short, including such legends as the Amazon women, Celtic women including Bodicea, and even 'A'isha, the wife of the Prophet Muhammad, who led the famed Battle of the Camel against 'Ali. In many culture groups today, women continue to serve not only in military support but as armed combatants.

Women's combat roles are not just a historical memory but a recent reality. For example, in Mozambique from 1954-1964, girls and young women participated side by side with men in a guerrilla insurgency against the Portuguese colonists. They not only undertook intelligence and support activities but actively engaged in armed combat—a role that posed challenges for the Portuguese military.[6] As the case of Mozambique illustrates, in certain military contexts, Marines need to be prepared for the legal and ethical issues that can arise from conflicts involving female combatants or in engaging with militaries that employ women in roles not permitted within U.S. forces.

Marines should also take into planning and operational consideration the possibility that in extreme situations, such as war, men and women may sometimes "step out of character" and take on "nontraditional roles." In the Algerian war (1954-1962), for example, women joined the resistance movement, carrying weapons under their robes and even engaging in combat. This, however, did not necessarily imply a change in attitude towards women's rights or roles, as the current situation in Algeria poignantly illustrates.[7]

There are two implications for Marines of "in extremis gender role alteration." First, crisis situations will influence the social roles, access, and behavior of men and women, changing them from the "traditional" roles with which Marines may be familiar, and permitting new kinds of interactions between men and women, to include Marines. Second, the change during extreme times often does not signal a sustained change to the over-arching social order with respect to gender. Here, what a Marine might see at one point during a deployment might not necessarily indicate that the social rules have changed.

Not only does gender affect the kinds of labor men and women are willing to perform, but in a particular AO, communities will follow an acknowledged or implicit consensus about how *everyday* participation in society should be gender-coded. Specific social activities are often considered "male" or "female" activities. In many Midwestern and Western U.S. states, for example, drinking alcohol in the local tavern is primarily a male activity. Similarly, many Americans view shopping to be a "female" activity. A man who hangs around the mall on a weekday would probably be considered strange and perhaps dangerous, just as a woman without an escort could expect to receive a lot of stares if she stopped in the local tavern in Ovando, Montana.

Example: Gender and Activity in a Small Area

In the weekly markets in southern Tunisia, there appears to the casual observer an undifferentiated jumble of males and females of all ages who sell, transport, and shop. However, there is a clear gender division of activities. Men generally bring the goods to market and do the selling. They move goods to market in cooperation with adult brothers, cousins, and perhaps brothers-in-law, while adolescent males often move the goods around the open air market precincts, and run to get food for the vendors.

Women do most of the shopping. In some regions, purchasing is also gender-differentiated—though not according to "male goods" and "female goods," but according to category of commodity, such as clothing vs. produce. Often a particular sales spot will be run by several men from the same family. Although not visible to the outsider, however, women also play an important role in preparing goods for the market. The bread sold by the man at the booth was probably baked by his wife; the pickled vegetables on the table were most likely grown and pickled by the women of the family; and women typically make the rugs and blankets for sale.

This situation offers several implications for Marines at the tactical level. A Marine patrol at the outskirts of a market such as these observing women milling about might determine that it is a space inappropriate for American males to enter, or that since there are mostly women, there would not be good opportunities for rapport-building and information/intelligence gathering. This would be an incorrect assumption.

Example: Gender and Activity in a Small Area; continued.

Conversely, detaining all the male vendors in a certain area of the market would economically injure the interests of specific geographic communities in the surrounding area, while leaving younger male family members who were not in the immediate vicinity at the time without a ride home, but with a bad attitude. It would also negatively impact any civil affairs efforts to develop rural cottage industries providing economic opportunities to females.

Along with general gender-coding of activities, certain kinds of formal organizations and informal groups will often be gender-segregated, and specific purposes will be served by organizations dominated by men or women. In much of Africa, for example, organizations to combat HIV are dominated by female community members and medical personnel. Understanding what formal and informal bodies are dominated by which gender for which purposes will help to smooth Marine interactions with an AO's populace.

In this respect, Marines also need to recognize the implications of placing a male or female in charge of a project or group that is predominantly populated by one gender. A male Marine will probably not have much success if assigned to head up a sewing co-operative in Thailand. Likewise, a female Marine may encounter serious resistance if appointed as the chief point of contact for a predominantly male run construction project in Bolivia.

Gender not only affects the work and social activities of women and men in a culture, but it may also influence the *spaces* that each sex may use. Across a large region ranging from West Africa to Central and South Asia, men and women are traditionally physically separated. Gender separation, often referred to as purdah or seclusion, describes the practice of dividing space into a female realm (typically indoors such as the home), and a male space (commonly an outdoor or outside space such as the streets or cafes).

In working in such gender segregated societies, it is important for Marines to understand which spaces are primarily "female" and therefore off-limits to male Marines in most circumstances. Similarly, female Marines working in gender segregated societies will find that their presence in "male spaces," such as street cafes in some Middle Eastern countries, will often be uncomfortable. The important point here is that gender segregation not only restricts women to certain spaces, but that as a result, men are also limited to the spaces that are acceptable for them to enter.

In most social structures, gender differentiation has implications for dress as well. In many gender segregated societies women will cover their heads and bodies when they move into outdoor male spaces such as the streets. This practice, which Westerners refer to as veiling, is very foreign to Americans today, who are accustomed to the free mixing of the sexes and very liberal dress codes.

There are many reasons why women cover their heads and bodies in other cultures, and a complete discussion is beyond the scope of this book. However, veiling is not necessarily imposed upon unwilling women by oppressive men nor is it necessarily a sign of religious fervor. A number of studies show, for example, that the recent resurgence of the *hijab* (a pan-Islamic head covering) appears to be a popular political and social statement by women of their Muslim non-Western identity, rather than an expression of deepening religious attitude.[8]

Covering hair, face, body, or hands can also be a practical way for Muslim women to affirm their modesty and virtue when working or studying with strange men. In other words, by putting on this new head covering, women are making the statement, "I am not a loose Western woman." Rather than symbolizing separation from the outside world on religious principle, this covering can actually facilitate modernist and more open interaction with the (male) public domain.[9]

This observation is borne out by recent field observations by the authors in Muslim Tunisia, where several young women wearing the *hijab* were observed walking past the mosque during Friday prayer eating ice creams. Not only were they *not praying* during one of the major prayer times in Islam, but they were openly displaying their disregard for Islamic norms and conventions. Overall, it is important to recognize that there are as many reasons behind the clothing choices of men and women in foreign areas as there are behind the (often bewildering) clothing decisions of Western males and females.

The division of labor, social activities, space, and dress according to gender may lead the Marine working in a foreign AO to conclude that in other societies women are not politically or socially significant. The absence of women from the streets in Muslim countries, the apparently subservient roles of women in many Asian countries, and the low participation of women in the formal economy or political systems in many countries seems to suggest that women have little power or influence in foreign regions. This, however, is a misconception.

In many regions—and in particular in areas such as the Mediterranean and Latin America, where concepts of male machismo are important—the public display of male dominance is often critical for a man's sense of masculinity. In such societies, a man's honor depends on his ability to demonstrate authority and control over his family and, in particular, over the sexual behavior of his women. Although it may appear to the outsider that women are passive and dependent upon males in these societies, cultural research challenges these images of powerless females.

It is important for a Marine to understand that in these cultural groups, the **public *display* of authority and power is critical for men.** However, outside of the public view, **in the home, women often play extremely powerful and influential roles.** In fact it is commonly accepted in many Arab and Latin American countries that while the man does the public negotiating and represents his family or kin group in formal occasions, agreements are often not final until he returns home and gets the support of the female members of his household.[10]

Since modern industrial economies require men to be away from home for the day (or even for months), women often take on the role of social networking. By maintaining critical economic, political, and social ties, women create a network of social support that serves as a source of information and power within the community, as well as a channel of access to critical economic resources.[11] As Marines working in Iraq have discovered, Muslim women may be key figures in illegal distribution networks. Equally important, these networks create a social web of obligation and exchange, holding groups together.

Example: Women and Networks

During his Relief-in-Place (RIP), a Marine commander in Iraq received a map of tribal lines of influence. It included only the males of the AO—women, he had been told, were of negligible social influence, and in any event should be considered "off-limits" to (mostly male) Marine units. This new commander was short on male troops, however, and so elected to use his female personnel outside the wire, for civil affairs patrols, searching of Iraqi females, and controlled cordon-and-knocks.

When he did this, his patrol debriefs combined with human intelligence leads began to provide an entirely different picture of family connections, lines of influence, alliances, and sources of sensitive, exploitable information about the community's leaders. The new map included which males were having extra-marital relationships with which females (who are also sisters, wives, and mothers of high-interest males); which males were impotent, sterile, or known deviants; which males, females, and families were disappointed with marriage alliances or felt cheated economically in the contraction of marriage alliances; and which men (and their women) were secretly supportive of the coalition presence or were covert allies in the struggle for democratization, but who could not speak out, either based on a campaign of murder and intimidation, or because of compromising information that fellow members of the community possessed about their personal lives.

This Marine commander was able to alter the actions of his patrols accordingly, tacitly assisting those supportive of the Marines, and using compromising information as a blocking or controlling mechanism. Returning to the United States, this Marine commander understood the integral role of women in all networks, even gender-differentiated ones.[12]

Related to the social networking role of women, in negotiating with "formal leaders" in countries that emphasize male machismo, Marines need to be aware of women's potentially important roles in backroom political negotiations. Due to the need to publicly display power and authority, men in such countries may be unable or unwilling to be seen negotiating and making compromises with rival tribal, religious or ethnic leaders. Not bound by such restrictions, women may serve as hidden go-betweens, making informal visits and proposing informal deals between rival groups.[13] Conversely, in some situations, a male ethic of magnanimity to "weaker" people can work to the advantage of negotiating or planning teams that possess appropriately "feminine" female Marines. There have been cases where foreign male interlocutors felt unable to say "no" to a U.S. female, or were predisposed to be more agreeable, either based on a gender-coded sense of honor, or on a (false) supposition of future personal benefits.[14]

As all of these examples illustrate, the critical issue here is not what men do or what women do—because so much of that is region-, personality-, and context-specific—but understanding the gendered, complementary, and symbiotic relationship between them. Thus, in planning for operations in a region, Marines need to assume that *both* genders will play meaningful roles that affect operations. Indeed, as seen above, to be successful (particularly in Muslim and Latin American cultures), Marines may find that including both males and females in their operational teams will allow them access to critical domains and activities that single sex teams cannot enter.

Culture Operator's Questions: Gender

- What work, roles, activities, and spaces are assigned predominantly to men and women?

- Who undertakes which tasks and where?

- How must operational plans change to account for different work, roles, and spaces assigned to men and women?

- What roles do women play in local militaries and insurgencies? Do they engage in armed combat?

- If women are not visibly observable, what roles and tasks do they undertake "behind the scenes?"

- How can operational plans and assignment of manpower include gender to maximize effectiveness of the unit?

Kinship and Tribal Membership. Virtually every culture group around the world identifies members as belonging to a family or kin group. In the United States, when we refer to family, most Americans mean the nuclear family.   However, even in the U.S. most people would consider their family to include extended relatives: grandparents, aunts and uncles, cousins, in-laws and so on.

In many non-Western culture groups, the concept of family is extended even further.   Distant relatives—such as the children of one's great grandfather or even the relatives of one's mother's brother—are considered important members of a person's kin group. In such cultures, an individual may view him or herself as being related to hundreds or even thousands of people. Languages reflect this phenomenon as well. From Morocco to Western China, languages have different words for extended family members— cousins, uncles, etc.—depending on what side of the family they are from and their generation.

Not only may other culture groups hold a wider definition of who is in one's family, but the way that people are considered "related" is culturally based. In the U.S. we consider family to be a biological concept: one has to be biologically descended from a common ancestor for people to be "related." In other culture groups, people can become related by undergoing special ceremonies (adoption is essentially the Western version of such ceremonies); by sharing trials or ordeals; by co-residence (someone lives in the household long enough that everyone assumes "Uncle Frank" is really an uncle); or even by sharing a cultural symbol (or totem) such as the Bear or Eagle. A number of Northwest Coast Native American groups, for example, count all people belonging to a specific clan, such as the Bear clan, as relatives. There may be no biological relationship between members of the Bear clan, but because they are considered relatives, members of the same clan may not marry.

Beyond this, in different cultures the meaning, or *value*, placed on being "related" to someone differs. We can call this the "the social valence of kin relationships." In many cultures, a brother-in-law is more important emotionally and in terms of moral compulsion than he is in the United States. At a more basic level, the significance of being a brother or sister, in terms of care and obligation, shifts from culture to culture, given size of families, age split among siblings, the degree of urbanization, legal conditions, and even education. In short, how much and in what way people care about relatives is variable across cultures. Finally, certain kinds of kin-based relationships have specific implications and culturally-based expectations.

Example: Cultural Expectations of Kinship Relationships

In Turkey and many Middle Eastern and North African countries, the relationship of a son to his mother is typically stronger and takes precedence over the relationship of a man to his wife—definitely in the eyes of the mother, and often in the eyes of the son as well. The daughter-in-law does very little for the mother except to compete for the affections of the son; by contrast, a son can ensure his mother's material wellbeing, emotional support, and social status due to his accomplishments. As a result, young wives frequently are placed in conflict with their mothers-in-law over control of their husbands/sons.[xv]

Historically, among the Navajo Indians, a woman's brother often played the major "father figure" to his sister's children. Husbands—who had ongoing economic and social relationships and obligations to their sisters and mothers—frequently spent more time with their family of birth, than with their wives and children. As a result, in traditional Navajo society, the father-child relationship was much weaker and far more distant than the uncle-niece/nephew relationship.[14]

Sometimes a kin group is held together by the idea that everyone is descended from one common ancestor—a mythical great-great-great-great- grandfather or grandmother. Groups that count their lineage from and even call themselves by the name of this great-great relative are sometimes called tribes. In the Middle East and Central Asia, for example, many people will have a personal name, a family name, and a tribal name—based on the name of a fictive or real common forebear.

It is important, however, to realize that simply because groups consider themselves descendants from a common ancestor, this does not mean they are tribes. Tribes have a number of critical features that separate them from other large kinship groups.

First, tribes must have a corporate identity; they must not only recognize that they share a common real (or fictive) ancestor, but also consider that their common lineage bonds them together as a functioning group.

Second, people must use this corporate identity as a structuring principle for their group. Individuals must be given a position and role within the tribal structure according to their place within the lineage, some lines of the tribe being considered of lower or higher status than other lines. **Third**, the group will have a formal leader or set of leaders, designated to speak for the group, who are selected, at least in part, on the basis of their inherited position within the tribe.[17]

In many countries, kinship and tribal membership are major factors in a person's ability to find a job, get a promotion, rise to a position of power or authority, or even to gain access to essential goods and services such as water or medical care. Not only may relatives offer important connections, but specific kin groups may also control access to important physical resources such as land, water holes, mountain passes, farmland or grazing areas, and even the right to raise certain crops or harvest trees such as coconut or date palms. As discussed in Chapter Three (Environment), land and water and other physical resources often become intimately connected to culture. When operating in a foreign AO, Marines need to be aware of the way that kinship relationships organize and control access to physical and social resources in the community.

Strangely enough, kin groups may also possess special rights to certain kinds of knowledge, skills or beliefs. In Greece and Italy, for example, the right to care for a religious site or shrine, or to perform religious duties, may be inherited from generation to generation within a family. And among the Navajo Indians, certain sacred healing knowledge is passed along kinship lines and restricted to members of shaman families.

In recent years, due to American engagements in Iraq and Afghanistan, tribal organization has become of central concern to the U.S. military. Much imprecision remains, however, in how we approach tribes. Particularly in crisis-burdened areas, Marines are unlikely to find tribal structures intact. Whereas U.S. personnel might expect to see this:

Figure 5.1
Tribes as They are Supposed to Be

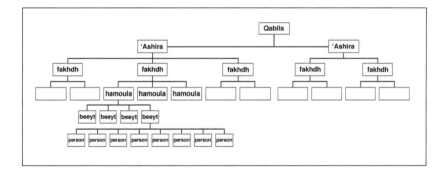

it is quite likely they will encounter something different, that looks rather like this:

Figure 5.2
Tribes as They Often Are

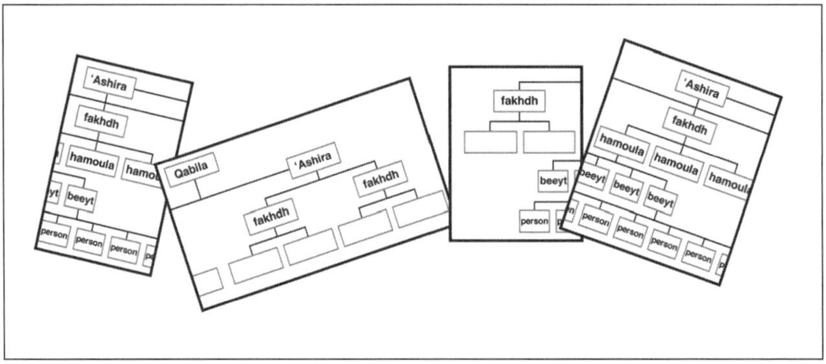

It is essential to avoid pre-conceived notions about tribes as social structures and nodes of social control, based either on romantic ideas, or one's last deployment.

Example: Tribes and Change

Iraqi tribes have been manhandled by the state for over a century. The Ottomans worked to sedentarize them, turning their sheikhs into absentee landholders, thus injuring the sheikh-tribe consultative relationship. The British then assumed sheikhs were like Victorian landed gentry. They exaggerated their waning social power, and used them as (often ineffective) nodes of authority and coercion in the countryside, while building certain ones up economically—separating them even more from their tribes. The Iraqi monarchy then used laws, schools, and conscription among other things to strip sheikhs—and tribes—of all power to get in between citizen and regime, and distanced sheikhs further from their regions by providing them parliamentary sinecures and stipends.

The policy of the post-1958 Republican governments was avowedly anti-tribe, since they considered them barriers to Iraqi citizens who would identify with regime and Arab nation. Though using Tikriti and Al Bu Nasr tribal connections to attain power and eliminate rivals, Saddam continued this broad policy of reducing tribes to political insignificance.

By the 1980s, this, as well as increasing rural-to-urban migration, had resulted in sheikhs at every level with much diminished power, and often a legitimacy deficit and eroding social esteem; fragmented tribal structures; and state-, religion-, or region-based identities.

Example: Tribes and Change; continued.

This reality of fragmented tribes and enervated tribal leaders has led some U.S. commentators and commanders to consider tribes in Iraq as "bad," "backwards," or simply no longer significant. However, this neglects two important points, one historical, the other conceptual.

First, after the Gulf War, seeing the need for some social support in Iraq, Saddam permitted people to use tribal names, and in areas where local leaders were overtly supportive of him and willing to exert social control on the Ba'th's behalf, he provided funds, empowered leaders, and permitted some legal autonomy to tribal leaders—many of whom he appointed. This was all supposed to create a neo-tribal structure with Saddam at the apex.

Second, in spite of the *reality* of fragmented tribes whose powers of social cohesion and control are much enervated, many Iraqis—even those who are detribalized, or in cities—*identify* themselves as members of tribes, and *identify* with tribal modes of support, power, and justice—either as necessary evils or as positive values.[18]

This all means three things. First, a history of tribes does not mean a Marine will wade ashore into a present tense of tribes and all that means for social power. **Second**, and related, although kinship and tribal affiliation are significant in some cultures, Marines should not assume that tribes are the only or even main factor in determining social structure and power around the world. Other cultural features such as ethnicity, gender, or class may be far more relevant than kinship in defining a person's status and power in a certain region, or in a certain situation. **Third**, decayed or dead tribes do not mean that a tribal identity or tribal norms have also gone away.

Beyond these immediate issues, due to our current experience with Afghanistan and Iraq, we tend to find "tribes" in regions that do not have tribes, but other kinds of kinship relationships. Recent military writing on the Philippines, for example, has begun to describe the system of political dominance by prominent families in terms of tribal identity. This is a serious mistake which obscures the significantly different yet equally important nature of a central oligarchy based on elite families.[19] Recent scholars have shown that same mistake has been made in post-1991 analyses of Central Asian political alliances.[20]

Marines also need to realize that due to marriage between kin groups, in any specific community a person may belong to more than one tribe or kin group. Hence "tribal affiliation" is not a fixed status but a fluid and changing concept. These cross linkages between kin groups can be especially significant in creating bridges of communication between the groups; the same cross-linkages, however, can place individuals in positions of conflict or jeopardy when alliances are tested or broken.

Culture Operator's Questions: Kinship and Tribal Membership

• How are land, water, or access to certain goods and resources concentrated in the hands of specific kin groups or tribes?

• How will our operations in the region support certain kin groups and enhance their power; or conversely undermine these groups?

• What are the possible outcomes of an operation that will challenge the power or control of resources by certain kin groups in the region (war, insurgency, increased stability, greater/lesser access to important goods and services)?

• How does a Marine's choice of local points of contact interact with or disturb local kin relationships, thus influencing the degree of success of Marine initiatives?

Class. Around the world almost every culture group organizes members on the basis of the three previous categories of age, gender and kinship. As societies become larger and more complex, they begin to stratify and distinguish among people through a fourth category: class. Class is a complex concept that is much more difficult for a Marine to recognize than age, gender or even kinship. This is because though one is often born into a class, this concept is not based on biology.

Although class is often related to wealth, other cultural factors such as education, one's occupation, parentage, region of origin, or even the way one acts and speaks may all be components of class. Class is culturally variable and the definition of class will vary from country to country. For example, in the U.S. one's class is largely based on one's wealth, and to a lesser degree on one's education and occupation (which are generally correlated with wealth).

In Europe, on the other hand, class continues to be defined by inheritance and proper breeding, and class definitions based on wealth are frowned upon. In England an aristocratic family that has "fallen upon hard times" is still considered upper class, even if the members cannot afford the lifestyles of the rich and famous. In contrast the *"nouveau riche"*—who have lots of money but no manners or breeding—are generally excluded from upper class British society regardless of how many yachts or castles they buy. For the purposes of this book we define class in a particular way.

Class: A way of stratifying groups of people according to their economic status and power in a society. Certain social characteristics such as the accumulation of goods or other forms of wealth; education; occupation; region of origin; lineage; and social behavior may all be indicators of class. However, since these indicators are culturally coded, class will be based on different characteristics in different societies.

This definition of class has three implications. **First**, class may be based on a single variable or a combination of factors. **Second,** a Marine cannot assume that his or her understanding of class differences in one culture will apply to another culture. In other words, what factors indicate "class" in one culture group may not indicate "class" in another—or have the same *weight* as a determinant of class level. Thus it is important to understand the AO first and avoid an *a priori* assumption of what class means. **Third,** the variables listed above may determine one's class; conversely, one's class may influence one's ability to access to some of the factors, such as education, occupation, and wealth.

An important feature of class systems is that they are based on a clear hierarchical stratification system. This stratification is directly related to the wealth, power, access to goods and privileges of the members of each class. At the top of the social structure are the privileged elite who control the wealth and political structure, while at the bottom are the poor and marginalized members of society who have virtually no power or control over the social structure.

Some class systems offer upward mobility to the lower members through education, hard work, or other incentives. Other class systems are rigid (for example the aristocracies of Europe) and do not permit movement or marriage between classes. Generally, societies whose members feel they can attain upward mobility from working *within* the system are more stable than societies with rigid class systems, whose members may perceive no opportunities for in-system mobility. In order to keep the lower classes silent, rigid class systems require stronger militaries, greater control of the wealth, education and social resources of society by the upper classes, and numerous social structures (especially religious structures) that support and justify control of society by a select elite.

Example: Class and Conflict

Many of the world's conflicts today seem to be related to ethnic or religious issues. In Europe, however, class has served as probably the greatest cause of recent conflict. The French Revolution was clearly a class based revolution, as the peasants and merchant class sought to overthrow the aristocracy. The American Revolution was also a conflict of class: colonists challenged the right of a distant aristocracy to rule and tax them and developed a Declaration of Independence founded upon the equality of all citizens. The American cultural dislike for class systems (although the U.S. does indeed have a class system) reflects these anti-class principles first set down in the Declaration. The Communist Revolution in Russia at the turn of the twentieth century was yet again a class conflict. In this case, however, the aristocracy was not only removed, but a clear anti-class philosophy called Communism was adopted. Bringing together class, ethnicity, and religion, in parts of the Balkans in the late nineteenth century, Greek and Slavic Christian nationalists hoped to agitate their agricultural peasant communities against Ottoman Muslim landholders and political overlords. Changing the names, similar dynamics uniting class, ethnicity, and religion could be seen a century later in this region.

Note that in our discussion of rigidity and stability in class systems, as well as in our example above, a key element is perception. A class system might in fact be rigid, not permitting extensive upward mobility—or there might be objective avenues for movement. More important, however, is the *perception* of disadvantaged sectors with respect to opportunities for mobility. Likewise, just as important in determining stability of a society is individual's and groups' judgment as to the possibility of *in-system*—as opposed to system-breaking—mobility.

Culture Operator's Questions: Class

• How is class defined in the AO: on the basis of wealth, education, region of origin, inheritance, or other factors?

• What are the privileges (economic, political, social, religious) of members of the upper class?

• How is access to essential resources for survival (food, shelter, clothing, water) determined by class?
• How does the concentration of wealth (through corruption, graft, or legitimate means) in the hands of an elite upper class relate to resource or power access?

• If creating a plan to support lower class groups, will funds and resources have to pass through the hands of the upper class first (and consequently disappear)?

• What is the reality of upward mobility in the AO's class system, and what do local people consider to be their potential for in-system upward mobility?

• How will Marine measures that influence different groups' social mobility be viewed by those groups, or by other, competing groups?

Ethnicity, Ethnic Identity, and Membership. Ethnicity is probably one of the most discussed and least understood forms of cultural identity and membership. In Chapter Seven we will discuss ethnicity as a form of identity. In this chapter, however, we will concentrate on ethnicity as a form of group membership. Although the symbolic and psychological dimensions of ethnic identity are extremely interesting, the Marine working in an AO requires a concrete, if simplified definition of ethnic identity. We provide one here to assist Marines in recognizing and identifying ethnic groups and differences fairly quickly upon arriving in a region.

Ethnicity: identification of an individual with a unique subgroup in a society, which is distinguished by specific behaviors, characteristics, and social symbols that can include, but do not require, a *language* specific to the group; *symbols* reflecting group membership or carrying hidden meaning for group members; unique *traditions*, rituals and holidays; clothing or *dress* unique to the group; a shared sense, or *memory* of history—often enshrined in mythical stories or folk tales; attachment to a *place* or region that holds symbolic meaning.

Although each of these are factors that may describe an ethnic group, Marines must keep in mind that not all ethnic groups will display all of the behaviors or characteristics above. Likewise, individuals in the group will vary significantly in the degree to which they actively reveal their ethnic membership to others. Although some members may speak only the language of the group, wear traditional dress and never leave their community of origin, others may live a "double life": wearing clothing typical of the majority group in the region, speaking the national language and working far away from their home community. Such ethnic members may express their identity only at special holidays or when they return to their village of origin.

In the same way, a Marine must remember that there is nothing specifically or scientifically "ethnic" about language use, historical memory, etc. What is important is that people who identify themselves with an ethnic group—or identify others as parts of ethnic groups—tend to use these characteristics to describe their or others' ethnic membership. It is exactly this malleability of ethnic identity that makes ethnicity difficult to pinpoint. For unlike gender or age, which are usually fixed identities, **ethnic identity is context specific,** expressed or hidden according to the political expediency of the situation.

Although in theory a person is born into a specific ethnic group and remains a part of that group for life, in reality, people have quite a lot of choice in the degree to which they affiliate with a specific ethnic group. Due to intermarriage, for instance, a person may belong to more than one ethnic group. It is not uncommon in Pakistan, for example, for a child to be raised in a household where several languages are spoken—Urdu by the Karachi-born father, Pashto by the Peshawar-born mother, and perhaps Persian by the grandmother whose family migrated from Iran—all reflecting the differing ethnic heritages of his or her parents. As an adult, he or she then has the option of claiming one or more ethnic affiliation—a choice that can be used to advantage in getting jobs, immigrating, or gaining political power.

For example, though growing up in Diyarbakir in eastern Anatolia and likely of Kurdish parentage, Ziya Gokalp chose Turkish as his identity and language, becoming one of the ideological fathers of modern Turkish nationalism in the late nineteenth and early twentieth centuries.[21]

Example: Ethnic Identity as Created and Malleable

The Israeli-Palestinian conflict is a curious case of the malleability of ethnic identity. People on both sides of this conflict created the idea of an ethnicity. They then found or created bases (linguistic, historic, territorial, etc.) for an identity related to that created ethnicity, and used that identity to restructure the local power structure.

On the Israeli side, there had never been an "Israeli nation," though there had been kingdoms of Israel and Judea, which were monarchies supported by a priestly caste. People living in these kingdoms were called Hebrews by outsiders, with the explicit understanding that their religion focused on a Temple cult in Jerusalem.

Example: Ethnic Identity as Created and Malleable; continued.

After destruction of the final Judean kingdom in 70 CE, there emerged a Diaspora religion of rabbinical Judaism. Only in the late nineteenth century, when secularized and sometimes even anti-religious German and Russian speaking Jews espoused a European-style nationalism, did Zionism emerge, as a reaction to European nationalisms and anti-Semitism.

Zionism is thus a modern nationalism of Diaspora Jews who used symbols of the religion, yet sought to create a secular identity focused on a re-cast notion of the strong Hebrew farmer, pioneer, and warrior—the antithesis of the Diaspora Jew's reality. Palestine was chosen as the territorial locus, where a political state could be built.

We thus have a pseudo-ethnic reconstruction of Judaism, at the same time as it was a rejection of the mainstream body of Jewish religious writing and belief, which saw return to Zion as an end-of-days messianic affair.

Notably, Zionism was also the effort of a particular social group—Central European white-collar urban intelligentsia—to overturn their status of social and political exclusion from both the traditional rabbinical hierarchy and the Christian mainstream, by redefining who they were, teaching other Jews to instead become Hebrews.

New Hebrews would speak a new language—partially recreated by German Zionists—instead of German, Russian, etc; they would have a "national flag" borrowing religious symbols, but with no historical pedigree; and they would remember "Hebrew" history and "Hebrew" holidays, in a way totally averse to traditional Jewish understandings. They would do all this in a land which none of their forebears had known for a millennium—while living in regions of that land outside major demographic centers of the kingdoms of old. The presence of another people—perceived as enemies rejecting national re-birth—furthered this pseudo-ethnic solidarity, based on traditional biblical or Jewish ideas, such as "Jacob vs. Esau," "the few against the many," etc., which were used now for modern political nationalism.

On the Palestinian side, though Arab cartography had long referred to a "Filistin"—as had the Romans and Byzantines before them—it cannot be said that the Christian and Muslim Arabic-speaking inhabitants of the Holy Land came into the twentieth century conscious of being a nation—an entirely European concept just infiltrating into Ottoman lands—nor were they conscious of a unifying Palestinianess. Rather, identity was universalist—Islamic/Ottoman—or hyper-local, based on local town or region (Nablus, Hebron, etc.), or religious sect.

Example: Ethnic Identity as Created and Malleable; continued

To the extent that there was another local solidarity, it is likely to have been Levantine—the Greater Syria *(Bilad al-Sham)* of Palestine, Lebanon, and Syria. Only after the first decade of the twentieth century did this begin to change. The encounter with Zionism—a self-consciously ideological, pseudo-ethnic, and linguistic nationalism—occurred at the same time as the educated Arabs of Palestine were influenced by Turkifying currents of thought and policy emerging from Ottoman Istanbul. The same currents had also spurred a growing, though minority body, of Arab nationalist thought and literature, which had influenced some Arabs of Palestine.

It should also be said that both Arabism and Palestinianism bore with them a certain social revolt among younger, better educated, and urbanizing Palestinian men, who were chafing at the dominance of the traditional grandees within Palestinian society.

It was only after the political dismemberment of *Bilad al-Sham* into British controlled Transjordan and Palestine and French-controlled Syria and Lebanon, and the emergence of embryonic states in all but Arab Palestine, that a sense of separate Palestinianess began to crystallize. This sense was furthered by conflict with the Zionists.

Complete Arab defeat in 1948; the removal of traditional Arab elites from the Palestinian scene; the attempts of Arab states to restrain the activities of the Palestinian refugees; and an Israeli campaign to erase any traces of Palestinianess from the new State of Israel combined to spur a defensive movement of Palestinian identity, politics, and violence.

Part of this was the creation of a particularly Palestinian notion of nationhood, re-writings of the "Palestinian" past, and sustained campaigns to establish a Palestinian political presence, both within remaining parts of the British mandate-era Palestine (the West Bank and Gaza Strip), as well as in Arab states and Europe. As with the Zionists, flags, anthems, particular memories of historical events, and the creation of heroes served to strengthen an ethnic identity of Palestinian national group membership— though to this day, it has not entirely disentangled itself from the circles of Arabism, Islam, and now, Islamism.

Modern ethnic and ethno-national identities across the world share much in common with that of the Zionists and Palestinians, as they are created in the midst of social and political conflict for specific aims, and can evolve over time, even in the midst of conflict—note the increasing Jewishness of Israeli identity over the past twenty-five years, along with the growing Islamicness of what it means to be a Palestinian.[22]

This malleability in ethnic expression adds up to **four responsibilities for Marines. First**, they must do their best to identify the different ethnic groups in their AO. Ethnic membership (and its correlate, religious identity—see below) is at the heart of many of the world's current conflicts, and understanding the various ethnic groups in a region is often central to operational success.   Not only must the Marine identify the groups that consider themselves ethnically distinct in the region, but he or she must also evaluate the interrelationships between the various groups: their exchanges, shared claims to resources, and actual or potential conflicts.

Second, the "malleability" of ethnicity means that people carry with them several potential ethnic affiliations. They select a particular affiliation according to specific opportunities and challenges they face. Therefore, the Marine must not think that "this person is only and always a member of that group." Recalling "I.M. Marine" from Chapter Two, one can also see that a Marine can be an American, Texan, southerner, or Marine—given the appropriate time, circumstance, and social need. Marines must thus be open to the idea that different ethnic or quasi-ethnic memberships are activated by people in an (often unconsciously) opportunistic fashion to improve positioning or access within a fluid social structure.

Third, this means that it is essential not to take a checklist approach whereby ethnic groups are identified—or ignored—based solely upon the factors indicated above. In many cases, ethnic identity is based upon the group's own view of itself, and its symbols and characteristics are based upon what it—or its opponents—choose. Likewise, for reasons we have seen here, color-coded maps with clear dividing lines may be of little help in understanding the AO confronting the Marines. This means better knowledge of the AO in the cultural and human sense must take the place of checklists.

Although it is easy to blame ethnic conflicts on "cultural differences," the reality is that ethnicity, like gender and age, often relegates members of an ethnic group to certain limited roles with an associated status and power within the larger society. At the heart of most ethnic conflicts then, is a challenge to the existing power and status structure—an issue that we will discuss in further detail in Chapter Six. Though it is about positioning in the social structure, ethnic identity thus becomes used as an ideological rallying cry to fight for a restructuring of the social order which would give the ethnic group more power, status or autonomy.

Fourth, Marines must understand that although maps of a region which block out the location of different ethnic groups are visually appealing and simple, as we indicated above, geography does not link to ethnicity in an obvious two-dimensional pattern. As mentioned above, many people belong to more than one ethnic group: a fact that is obscured by two-dimensional maps. Likewise, many people have a physical place of residence, but also a spiritual or psychological "home" which may not match where they live. The Armenians outside of Armenia would be a case in point, as are even non-Zionist, but traditionally-minded Jews. Particularly for groups that have been expelled from their "homeland" this psychological home has strong meaning—leading often to decades of conflict or war; yet military maps fail to capture this concept.

Additionally, humans move, and national lines also move over time. Although the world is divided up according to countries with official borders, ethnic groups may be split across these borders. As a result, in many states, people are loyal to groups across the border: citizenship is no obstacle to loyalty to ethnically similar groups in other countries. This means that groups can be members of the social structure of more than one state, occupying different positions of prestige or influence in the two states.

Example: Colonial Powers' Maps and Ethno-Social Realities

Colonial powers often draw maps of their territories ill-attuned to ethnic groups and the social structures of which they are a part. Sometimes this occurs through ignorance of local dynamics or in order to placate local benefactors during decolonization. One of these cases is Ogaden, currently in Ethiopia. This area is ethnically Somali, and religiously Muslim, at odds with the majority social and political order of Ethiopia (Ethiopian Christians). Ogaden only came under Ethiopian political control at the end of the nineteenth century, when Britain was unprepared to venture beyond British Somaliland.

Example: Colonial Powers' Maps and Ethno-Social Realities; continued.

However, after the Italian fascists lost their portion of Somaliland to Britain, the latter moved to take over the Ethiopian-controlled Ogaden. Ultimately, Britain failed, so that Muslim, Somali Ogaden has remained part of Ethiopia, while Somalia has been a rival of Ethiopia under successive governments.

Over the past thirty years, Ogaden Somalis have often raised the banner of secession, at times helped by the Somali government across the border. In this case, Somalis in Somalia consider their Ogaden cousins part of their social structure and group; governments in Mogadishu have attempted to use this feeling to counter Ethiopian meddling in Somalia, and Ogadenians consider themselves part of the social structure of Somalia, though they seek an independent state through continuing insurgent actions.[23]

At other times, colonial powers intentionally create borders out of sync with local ethnic dynamics—particularly when the colonized areas are included into the expanding state itself, which hopes to prevent coherent social units that could be politically potent. The Soviet Union regularly gerrymandered internal republics' administrative borders to counter nationalism. Parts of Central Asian states today either contain pockets of people from different ethno-linguistic groups, or are surrounded by other states' territory.

For example, Samarkand, part of Turkic Uzbekistan, is in fact historically ethnically Tajik and linguistically Persianate. The city has been the site of tensions between the two groups, thus prolonging until today the Soviet policy of weakening both national movements. Likewise, while the Czars and Soviets exported Russians to Kazakhstan, at the end of the Soviet era, thousands of them moved back across the border, fostering a managerial and technical brain drain from the new state (the more than one million Kazakhs in China seem to just be an accident of imperial expansion).[24]

Although we think of a world of nation-states, most places where Marines operate have people who exhibit sub-national loyalties to something other than the state—down to the hyper-local level. In these cases, group membership, or ethnicity as a vehicle to jockey for position in the social structure, can be subdivided even down to the sub-company level. It is for all these reasons that the term "human terrain," though so congruent with military ways of thinking about tactical and operational matters, is, as we saw in Chapter Two, "static with respect to change over time; rigid with respect to fluid human relationships; and limited to representing behavior in only two dimensions."

Culture Operator's Questions: Ethnicity

In order to include ethnicity in military planning, Marines need to consider the following questions:

- What is the relationship between particular ethnic groups and control of professions or positions of power?

- How do groups that are barred from these positions of power challenge the system (breeding grounds for insurgents, theft and bribery, civil war)?

- What are local assumptions about U.S. and western biases and partisanship with respect to ethnic group struggles?

- How will a Marine alliance or dealings with a particular ethnic group affect those in power?

- What are possible reactions of those groups that are ignored?

- In this AO, what kinds of processes have historically activated which ethnic identities and feelings of group membership?

Religious Membership. In many ways, religious membership can be considered a special form of ethnicity—in fact, often those who affiliate with religions as social communities may have difficulty in practice distinguishing their religious membership from something akin to ethnicity. Religious membership, here, needs to be distinguished from religious belief systems. It is entirely possible for someone to hold a specific religious belief system and never belong to a religious community. And, although in theory this should not happen, there are many people who belong to a religious community, attend church or other formal religious activities, and yet have little personal faith or connection to the belief system. Therefore we define **religious membership** as being part of a group of people that considers itself united by religious faith.

Frequently membership in a religious community is demonstrated by many of the same markers that distinguish ethnic groups. People in a specific religious group may wear specific forms of dress, for example the saffron robes of Buddhist monks. Members will share special traditions and rituals that they celebrate at specific times of the year, such as Christmas and Easter for Christians. They may speak and study a unique language such as Aramaic for Chaldean Christians. And often religious groups have a spiritual connection to a physical place, such as Mecca for Muslims.

Finally, like ethnic groups, members of a religious group may be limited to undertaking certain jobs or tasks within the larger society. Interestingly, long after economic restrictions have been removed from the religious group, they may continue to work in certain occupations which they consider to be more "religiously identified" with their group.

As with ethnic groups, religious groups have vied throughout history for political power and influence. We are used to seeing these conflicts as outright war; as in the Reformation Wars in Europe and the post-1992 hostilities in the Balkans; as sectarian strife of an intermittent kind, as between Sunnis and Shi'is in Pakistan; or subnational sectarian war, as between Sunnis and Shi'is in Iraq. However, less obvious but just as important, political power plays between competing religious groups of the same sect are worthy of note, as they can cut through economic or government bodies in much of the world where Marines will deploy.

Example: Legitimate Religious Groups in Competition

In Western and Sahelian Africa, mystical religious brotherhoods abound. These brotherhoods are outside official Islam but still legitimate in the eyes of Sunnism. Some of these orders have traditionally been socially prominent, with members occupying important positions in politics and economics. These Sufi orders are thus inclined to competition for social prestige, based on the status of their members. In Senegal, the Tijaniyya and Mouride Sufi orders continually jockey for social influence. This competition spans politics, economics, and military authority, with the prestige of one or the other orders increasing as their representatives attain power, to include the presidency.

Culture Operator's Questions: Religious Membership

When creating an operational plan in a region where diverse religious groups are in conflict, Marines should look past the religious rhetoric to examine the economic, territorial, political, and power motives that may fuel the religious groups' agendas:

- How do people define and express their religious membership in the region?

- What roles and status do the various religious groups or sects hold in the larger society?

- What is the meaning of geography for religious groups in an AO?

- What effects would a planned Marine operation in the region have upon the power, status, and access to critical resources of the various religious groups or sects?

- How will the Marine operations influence indigenous peoples' views of Marine or U.S. biases towards different religious groups of the social structure?

Notes

[1] Robert Layton, *An Introduction To Theory in Anthropology* (Cambridge, UK: Cambridge University Press, 1997).
[2] See Setha M. Low and Denise Lawrence-Zuaniqa, eds., *The Anthropology of Space and Place: Locating Culture* (Boston, MA: Blackwell Publishing Limited, 2003).
[3] Amir Boktor, *School and Society in the Valley of the Nile* (Cairo: Elias' Modern Press, 1936); idem, *The Development and Expansion of Education in the U.A.R.* (Cairo: AUC Press, 1963).
[4] "Debrief: 2nd BN 7th MAR CAP Plt," TCMEF, Nov 2004, 29 Palms MCB.
[5] For this phenomenon in different parts of Africa, see Deborah Durham, "Youth and the Social Imagination in Africa: Introduction to Parts 1 and 2," *Anthropological Quarterly* 73:3 (July 2000), 113-120; Susan Rasmussen, "Between Several Worlds: Images of Youth and Age in Tuareg Popular Performances," *Anthropological Quarterly* 73:3 (July 2000); P.T.W. Baxter and U. Almagor, eds., *Age, Generation, and Time: Some Features of East African*

Age Organizations (New York: St Martin's Press, 1978).

[6] Harry G. West, "Girls With Guns: Narrating the Experience of War of Frelimo's "Female Detachment" *Anthropological Quarterly* 73:4 (2000): 180-194.

[7] Susan Slyomovics, "Hassiba Ben Bouali, If You Could See Our Algeria: Women and Public Space in Algeria," *Middle East Report* (January-February 1995): 49-54.

[8] See, for example, Arlene Elowe MacLeod, *Accommodating Protest: Working Women, The New Veiling, and Change in Cairo* (New York: Columbia University Press, 1991).

[9] Sharifa Zuhur, *Revealing Reveiling: Islamist Gender Ideology in Contemporary Egypt* (Albany, NY: State University of New York Press, 1992).

[10] See, for example, Evelyn Early, *Baladi Women of Cairo: Playing With an Egg and Stone* (London: Lynne Riener, 1993); Suad Joseph, "The Neighborhood Street in Lebanon," in Lois Beel and N. Keddie, eds., *Women in the Muslim World* (Cambridge, MA: Harvard University Press, 1978), p. 541-558; idem, "Working Class Women's Networks in a Sectarian State: A Political Paradox," *American Ethnologist* 10:1 (1983): 1-23; Diane Singerman, "Where Has All the Power Gone? Women in Politics in Popular Quarters of Cairo," in *Reconstructing Gender in the Middle East*, eds. Fatma M. Göçek and Shira Balaghi (New York: Columbia University Press, 1994), p. 174-200.

[11] See, for example, Early, "Baladi Women of Cairo;" Singerman, "Where Has All the Power Gone?"; Paula Holmes-Eber, *Daughters of Tunis: Women, Family, and Networks in a Muslim City* (Boulder, Co: Westview Press, 2003).

[12] Anonymous Members of I MEF Headquarters Group, Interview by Dr. Barak Salmoni, Al Anbar Province, Iraq, October 2006.

[13] Joseph, "The Neighborhood Street in Lebanon"; Idem., "Working Class Women's Networks in a Sectarian State."

[14] Interview with anonymous CAG Team Leader;   3rd CAG Cultural Debrief, TCMEF, Sept 2004.

[15] Jenny White, *Money Makes us Relatives* (Austin, TX: University of Texas Press, 1994), 70-86; and Margaret Meriwether, *The Kin Who Count: Family and Society in Ottoman Aleppo* (Austin, TX: University of Texas Press, 1999).

[16] Clyde Kluckhohn, Dorothea Leighton, Lucy Wales Kluckhohn, and Richard Kluckhohn, *The Navajo* (Cambridge, MA: Harvard University Press, 1992).

[17] Garrick Bailey and James Peoples, *Introduction to Cultural Anthropology* (Belmont, CA: Wadsworth Publishing Co., 1999).

[18] Toby Dodge, *Inventing Iraq: The Failure of Nation Building and a History Denied* (New York: Columbia University Press, 2005); Amatzia Baram, "Neo-Tribalism in Iraq: Saddam Hussein's Tribal Policies, 1991-96," *International Journal of Middle East Studies* 29:1 (Feb. 1997): 1-31; Faleh A. Jabar, "Sheikhs and Ideologues: Deconstruction and Reconstruction of Tribes under Patrimonial Totalitarianism in Iraq, 1968-1998," and Hosham Dawood, "The 'State-ization' of the Tribe and the Tribalization of the State: the Case of Iraq," both in *Tribes and Power: Nationalism and Ethnicity in the Middle*

East, edited by Faleh Abdul-Jabar and Hosham Dawood (London: Saqi, 2003); Peter Sluglett and Marion Farouk-Sluglett, "The Transformation of Land Tenure and Rural Social Structure in Central and Southern Iraq, c. 1870-1958," *International Journal of Middle East Studies* 15:4 (1983), 491-505.

[19] Mina Roces, "Kinship Politics in Post-War Philippines: The Lopez Family, 1945-1989," *Modern Asian Studies* 34:1 (Feb. 2000): 181-221.

[20] Adeeb Khalid, *Islam after Communism: Religion and Politics in Central Asia* (Berkeley, CA: University of California Press, 2007).

[21] Niyazi Berkes, *The Development of Secularism in Turkey* (London: Routledge, 1999); Uriel Heyd, *Foundations of Turkish Nationalism: The Life and Teachings of Ziya Gokalp* (New York: Hyperion Press, 1979).

[22] Meir Litvak, "A Palestinian Past: National Construction and Reconstruction," *Memory and History*, 6:2 (Fall/Winter 1994), 24-56; Tarif Khalidi, "Palestinian Historiography: 1900-1948," *Journal of Palestine Studies*, 10:3 (Spring 1981), 58-70; Beshara Doumani, "Rediscovering Ottoman Palestine: Writing Palestinians into History," *Journal of Palestine Studies*, 21:2 (Winter 1992), 5-28; Yael Zerubavel, *Recovered Roots: Collective Memory and the Making of Israeli National Tradition* (Chicago: University of Chicago Press, 1997); Rashid Khalidi, *Palestinian Identity* (New York: Columbia University Press, 1998); Gershon Gorenberg, *The End of Days: Fundamentalism and the Struggle for the Temple Mount* (New York: Oxford University Press, 2002); Aviezer Ravitztky, *Messianism, Zionism, and Jewish Religious Radicalism*, trans. Michael Swirsky and Jonathan Chipman (Chicago: University Of Chicago Press, 1996); Neville Mandel, *The Arabs and Zionism Before World War I* (Berkeley, CA: University of California Press, 1980); Meron Benvenisti, *Sacred Landscape: The Buried History of the Holy Land Since 1948* (Berkeley, CA: University of California Press, 2002).

[23] Bahru Zewde, *A History of Modern Ethiopia* (London: James Currey, 1991).

[24] Hugh Pope, *Sons of the Conquerors: The Rise of the Turkic World* (New York: Overlook TP, 2006); Rob Johnson, *Oil, Islam, and Conflict: Central Asia since 1945* (London: Reaktion Books, 2007).

Chapter 6

Dimension IV
Political Structure

We have seen in Part II of *Operational Culture for the Warfighter* that cultural dimensions are related to each other. In particular, Chapter Four showed that *environmental* factors produce *economic* structures, while Chapter Five demonstrated that economic dynamics determine the *social* structure—or status, position, and relationships among people and groups—in a society. In this chapter we turn to investigate *political structures* in cultural terms, allowing us to see how they are the product of a culture group's social structure.

Simply put, **economic and social structures shape the distribution of power and authority in a group. The map of that power and authority reveals a group's political structure, expressed in political mechanisms and relationships.**

We **define political structure** as:

- The way that power and leadership are apportioned to people, and exercised, according to the social structure of the society.

This suggests that for the culture operator, "politics" is not the study of presidents and political groups. Rather, it focuses on something more nuanced:

- The way that a culture group determines who has power and control (who is a leader and what admired traits of a leader are).

- How that power is legitimized and exerted over whom.

147

- How and why conflicts over power and control emerge within and between groups.

A central term of reference in this chapter is "power," which should be distinguished from "authority."

- **Power** is the ability to *control or influence* the behavior of individuals or groups of people.

- **Authority** is the legal or popularly granted *permission* to exercise power. It is *legitimacy* in the exercise of power.

For the Marine, **two caveats** are in order here. **First**, one needs to remember that authority does not always equal power—one who holds authority does not always have the *means* to exercise power. For example, south of the Texas border with Mexico, there are districts where the legitimate authority held by the Mexican police in no way translates into the power to control illegal movement of people and narcotics.[1]

Second, in mapping the dynamics of power and authority in a society, it is important to distinguish between formal political structures—often created by an *outside* force, such as an occupying military—and local political organizations developed from *within* the cultural group. Externally-imposed power structures are often viewed as illegitimate by local peoples and can result in a "weak state" with little *effective* power or authority over local communities. Likewise, in such situations, formal, or "modern" political structures can be animated by informal, "traditional" practices and norms reflecting existing social structures.

The British and Soviet experiences in Afghanistan bear this out. When British forces installed Shah Shuja on the "Afghan throne" in 1839, he possessed absolutely no legitimate authority. Likewise, the Najibullah government installed by the Soviets in the mid-1980s possessed no legitimacy. When the Soviets withdrew from Afghanistan, Najibullah's regime crumbled as rival factions fought for power in regions of the country, based on political alliances emerging from social structures of kin networks.[2i]

Our discussions, therefore, will focus on indigenous, culturally generated political structures in communities, rather than formal and often artificial states. In this context, the following facets of a political structure are most significant:

- The cultural control of power and resources: political organization

- Who holds the power: cultural forms of leadership

- Conflicts over power: challenges to existing political structures

For a Marine operating in a new AO, understanding a) the influence of economy and social structure on politics; b) the existing power and authority structure; and c) how to work with the individuals within that structure are critical to operational success.

Political Organization

The various models of political organization offered by anthropologists share a common theme: the political *structures* and political *norms* of cultural groups can be characterized by their degree of egalitarianism versus stratification.[3] One can thus consider political structures and norms as constituting a spectrum, based on the following definitions:

Egalitarianism: Resources, power, and decision-making are not concentrated in the hands of any specific individual or sub-group, but are spread relatively evenly across members of the group.

Stratification: Resources, power, and decision-making are limited to certain categories of people within a community, based upon their—or their group's—status, entitlement, and rank. Society is thus multi-leveled, with groups enfranchised or disenfranchised according to their place in lower or higher social levels.

Egalitarian   Highly stratified

←――――――――――――――――――――――――――――――→

Bands   Tribes   Chiefdoms   States

At one end of the continuum are cultural groups that share power and authority more or less equally, and are highly flexible in their leadership structures. At the other end of the continuum are highly stratified cultural groups that restrict access to goods and resources to certain groups.

The above diagram includes four kinds of political structures. In an operational culture sense, not only does each category possess a distinguishing set of features as regards structure and function, but more importantly, they constitute the over-arching categories encompassing many of the political arrangements which Marines will encounter. These four structures include bands, tribes, chiefdoms, and states.

In examining the following definitions, bear in mind **three caveats. First,** these descriptions of political structures are archetypal models, and the political structures flowing from them rarely exist in such a "pure" form. **Second**, none of these political structures is hermetically sealed off from the preceding one. At each point across the spectrum, structures will bear some features of each other. Some tribes look like bands, some chiefdoms look like tribes, and some states conduct themselves more like chiefdoms, etc. Like we saw with "formal" and "informal" economies in Chapter Four, there are also informal political practices and patterns of behavior. And, just as we saw that formal economies are enmeshed with informal economies, so too, the "formal" political structure of a state can be animated by "informal" practices associated with tribes or chiefdoms. Tribes or chiefdoms themselves are not informal or illegitimate; rather, in a state context, practices associated with them are often considered so.[4]

Third, although the political structures of different culture groups can be placed along a continuum, the Marine should not assume that this model means that societies progress through one stage to the next. It is not necessarily the case, for example, that a band will metamorphose into a tribe; in fact, many bands are simply absorbed into state societies. By contrast, it appears that the surge in tribal groups in the past half century is actually the result of the disintegration of larger state organizations and the failure of centralized colonial power in certain regions.[5] As states fail, governance falls back on local communities which turn to pre-existing kinship and social structures to determine leadership and authority—all with implications for egalitarianism and stratification within society. There is no "first comes this, then comes that" in the life of political structures.[6]

Political Structures Defined

Bands. A band is a small group of people who all know each other face-to-face, and work closely together for a unified purpose of survival. Their dependence on each other to reach a goal requires cooperative work, thus militating against hierarchies beyond those of age and gender. As a result of this egalitarian social structure, leadership is determined on the basis of skill. Leaders emerge in response to specific challenges or opportunities, and decisions are often made by a council of more respected band members.

From this perspective, groups operating in the initial stages of an insurgency often reflect the characteristics of a band. It is exceedingly difficult to profile them in terms of command and control structures. It is just as difficult to shut them down through decapitation strikes—if leadership is event-specific and episodic, the group can always live to fight another day, unless everyone is neutralized.

Tribes. The Marine who has worked in Iraq or Afghanistan might be surprised to note that tribes are considered relatively egalitarian in their distribution of political power. As we saw in Chapter Five, a tribe is a group based on the internal assumption of kinship. This kinship binds members together as a functioning group, in terms of economic, political, and social relationships. The *feeling* of kinship solidarity imposes on people these relationships and obligations. The kinship-based political structure creates lines of authority and decision-making in which leadership is determined by inheritance of a lineage-based position of power. For example, a close (male) descendant of a tribal leader usually succeeds the leader.

While in theory power is position-based, in fact, power will flow inter-generationally through competition within a small core group of people related to the previous leader, who have the legitimacy to contend for leadership. For example, though a tribal leader's first son is "supposed" to succeed him, his lower intelligence, or the greater popularity and economic success of the second son, may render the latter more suitable for leadership.

If we look at the way a tribe functions as a corporate group that looks out for the welfare of its members, we realize that most tribes have a fairly simple tiered structure, with leadership shared by (formal or informal) councils. As we saw in our discussion of communal distribution in Chapter Four, there is a strong expectation that goods and services will be distributed through the tribe—those who acquire goods will look out for those under them. In fact, in the absence of external destabilizing factors—such as political manipulation or interaction with the global economy—the degree of social and economic differentiation between the poorest and richest members in a tribe is generally very small in comparison to the extreme differences of wealth that occur in state societies.

Tribal structures also tend to have a "checks and balances" system in which several tribes or sub-tribal units in a region all maintain an uneasy balance of power and control. While one tribe may gain the ascendancy for a period of time, no one group gains complete control over the others. This tense, conflict-prone equilibrium is a clear contrast to state systems where power is concentrated in a central node of leadership.

Of course, this equilibrium is often subjected to external manipulation, by outside political regimes, by leaders whose legitimacy among tribal members is based on alternate belief systems, or by the mechanisms of the global economy.[7]

Chiefdoms. In contrast to bands and tribes, chiefdoms are political structures with a centralized chief, who possesses a subordinate council of advisors and functionaries—unlike a tribal leader's subordinates or council members who are lineage-based and do not have specific, stabilized "billets." Rather, in a chiefdom, people have official, named positions. To a great extent, these subordinates' power is based on their "billet," at least as much as on their name.

Chiefdoms also centralize authority, decision making, and administrative structures. They do not exhibit egalitarian political structures—they are stratified, and "first among equals" is anathema. However, unlike states which are backed by powerful militaries, chiefdoms must rely on the support of subordinate chiefs and groups, often through patron-client relationships, in order to legitimize the role of the chief. Tributes from subordinates, and exchanges of goods, wives, and land with the subordinate leader cadre, are essential to maintain the power and position of the leader.

If the chief is unable to maintain these exchanges of goodwill or favors, one or more contenders for the chiefdom are likely to challenge the authority of the chief and attempt to seize the throne. As in all patron-client relationships, the patron needs to have clients, and keep them happy, as much as the client needs the patron, and needs to serve his interests.

As such, a fair amount of the chief's effort is devoted to ensuring the loyalty of subordinate leaders, as his position is maintained through his own personal charisma and ability to provide or deny material welfare and social success. Chiefdoms thus have tension built into them, with constant threat of snowballing destabilization. Not only the Taliban, but other "warlords" in Afghanistan, exemplify chiefdom characteristics.

States. In anthropological and operational culture terms, a state is characterized by centralized authority and control, possesses defined territorial boundaries, exerts power through security forces, and controls access to resources.

To elaborate, **first**, states possess a highly centralized government over which there is no contention. This means that while people might disagree over how good the ruler is, no one is under the impression that he is not the ruler. Also, the state may not be effective, but no one questions the existence of the state, or that it is the state which needs to be effective. In 1789, French revolutionaries judged Louis XVI to be inadequate, but they did not question the existence of France, or that the state should be the dominant political player on French territory.

Second, states also have defined territorial parameters—states are not "moveable feasts" as bands, tribes, or chiefdoms may be. **Third**, states possess regularized security structures (military, police, etc.) that are loyal to the state itself, and not something outside the state's boundaries. Militaries might make coups against the regime, but they do not revolt against the idea of the state itself. In fact, in states, the security forces are the major legitimizing tool for the state to effectively assure its power over the people.

Fourth, states monitor and control the allocation of important physical and social resources. These include elements of the environment, such as food and water, as well as aspects of the economy, to include finances, currency, transportation, and communications. Social resources monitored, regulated, or controlled by the state include education, medical services, social welfare mechanisms and—sometimes—religion, either as doctrine, hierarchies, or public expression. Control of these physical and social resources is often linked to the perception of a state's legitimacy—in the eyes of the people, government, and even the international community. Hence a state's concern for its formal economy, as seen in Chapter Four.

There are several factors related to states. From the outset, in anthropological terms it is essential not to confuse regime with state. A regime is a ruling government. While some regimes may portray themselves as representing or being the state, a regime is no more than the current group ruling a country. A Republican-controlled U.S. administration is not the United States; it is merely the government of the day.

Further, regardless of terms commonly used synonymously in the United States, states are not nations. A state is a political structure; a nation, however, is a socially-created community based on an ideology and a claim to certain territories and rights. A "nation" is an identity. There might be several nations within one state. For example, think of Arabs and Jews in Israel; Kurds and Turks in Turkey; or Serbians, Bosnians, Croats, Montenegrins, and Macedonians in pre-1991 Yugoslavia. Conversely, not everyone in the state might necessarily view a particular nation as an objective reality (Turks considering Kurds)—though no one debates the existence of a state.

Finally, while we have been considering bands, tribes, chiefdoms, and states separately, in many parts of the world—particularly in regions to which Marines deploy—all these political structures co-exist, even in areas considered by outsiders and local regimes as states. What makes these kinds of operations so complicated at the squad-through-MEF level is that areas where Marines operate have so many kinds of political structures vying with each other at the same time. Also, while Marines see a particular overt structure— such as a state with a parliament and national leader—this formal structure likely masks the functioning of other informal structures upon which "state" leaders rely in order to have power. If this were not the case, Marines would not be there!

Who Holds Power: Cultural Forms of Leadership

Western states' citizens are accustomed to a hierarchical political system run by a clear formalized leader. Yet many societies around the world do not have centralized leadership systems. Indeed, they may not have an obvious leader at all. Furthermore, even if a formal leader is designated, he or she may not be the actual *effective* leader—in the sense of the person who gets things done or issues directions for people to follow. Instead, someone else in the group may serve as the real informal leader, while the formal leader is merely a figurehead. An obvious example of the latter is the royal sovereign in the United Kingdom, who has formal figurehead authority, yet lacks the effective authority of the Prime Minister.

Marines should thus begin with the realization that often there is a distinct difference between the person who is designated to lead versus the person who actually leads.

Formal Leaders. These kinds of leaders receive official recognition by the political structure and community. Often, formal leaders have titles such as "mayor" or "colonel" or "police chief" and may wear symbols (badges, special clothing, etc.) indicating their status within the community (see Chapter Seven). Typically formal leaders have special offices or places to receive guests and receive various legitimate financial and other regularized benefits from their position. As we saw in Chapter Five, part of this is due to their and their position's status in the social structure.

In political terms, formal leaders can claim the power that comes with their position regardless of their knowledge, background, or skills at accomplishing their duties—the system made them the leader, and their authority comes from their position. However, depending on the personality of the person in the position, and the political structure's strength, formal leaders may not actually have much control over the community.

Informal Leaders. Contrasted to formal leaders, informal leaders may not carry titles and symbols indicating their status, and their offices and spaces may not obviously indicate a person of power to an outsider. Or they may hold a formal position or status that is significantly lower or different from their actual authority and power. However, despite their lack of official trappings, informal leaders may carry more power and have more influence over the community than formal leaders.

Informal leaders may gain status through working with people and undertaking their leadership tasks. Alternatively, some informal leaders gain influence in a formal system because the community already recognizes their status, prestige, or skill. As a result, they are generally highly respected by members of the community. And, as any Marine who has been around quality SNCOs and CWOs will attest, typically in order to "get anything done," one needs to have the approval and support of the informal leader(s).

In the previous section, we examined four main forms of political structure. These structures are related to different forms of leadership that Marines need to look for when they enter a new AO. Like political structures, some forms of leadership will overlap or be appropriate to more than one political structure. Thus, instead of an "if this, it must be that" approach to grasping political structures and their relationship to leadership forms, it will benefit the Marine to embrace a flexible "if this, then it might be a little bit of that and the other" analytical method.

Marines should understand their AO in terms of these cultural forms of political leadership:

Leadership in Acephalous Societies. Acephalous (literally meaning: "without-a-head") societies are groups that have no one formal designated leader. As we saw above, many bands, for example, have no obvious designated leader. Since such groups are small (ranging from 10-25 individuals) most social behavior and decision making is communal or made along lines of gender or age.

A Marine who tries to negotiate with the "leader" of an acephalous cultural group is most likely to discover that any agreements made with this "leader" are unlikely to be recognized by the rest of the group. For example, negotiating with the self-described "leader" of an emergent insurgent or criminal gang will produce "agreements" to which neither the "leader" nor the gang feels obligated.

Episodic leaders. Episodic leaders can be found in all political structures, whether bands, tribes, chiefdoms or states. These leaders arise to undertake a specific purpose, whether to lead an attack on an opposing group or to run the PTA's soccer fundraising event. Episodic leaders have no formal official control or power other than that which is given to them for undertaking the specific task at hand. Once the goal has been achieved, episodic leaders step down to continue on with the normal routine of their daily lives.

Episodic leaders are not foreign to the U.S. military. Ad hoc task force commanders, for example, might be battalion operations officers or even company commanders. After they have completed the specific mission for which the task force was assembled, their attachments will return to parent units, and the task force commander will reassume lesser, regular duties.

Councils and Oligarchies.   Councils and oligarchies (rulership by an elite group) are extremely common forms of leadership for tribal groups. They are also fairly common in states and chiefdoms. The unique feature of a council is that no individual bears the right to make a final decision or to implement a course of action. The decision must be considered and shared among all leaders. During the conflicts between Native Americans and the U.S. military, failure to understand the importance of tribal councils led to many misunderstandings. In some situations, U.S. military leaders negotiated treaties with individuals who had no right to make individual decisions for their tribe. Such inappropriate negotiations led to invalid treaties and, in some cases, unexpected warfare.

As this example shows, councils and oligarchies may share characteristics in common with acephalous or episodic leadership systems when it comes to their interaction with external entities. However, what distinguishes acephalous and episodic forms from councils and oligarchies is the permanency and uncontested authority of the latter, as well as their regularized meetings, deliberations, and decision-making mechanisms. For example, though the Roman Republic had rotating consuls, the system itself was not episodic.

Hereditary Leadership.   For much of the history of the world, leadership has been inherited along family lines. Since a family, by definition, is generally of a common ethnicity, religion, socio-economic class, and regional origin, by default, hereditary leadership concentrates power in the same ethnic, religious, socio-economic, and regional groups. Hereditary leadership therefore discourages political mobility by reinforcing existing social structures and stratification: The same groups continue to be included in decision making, while others are excluded generation after generation.

Since hereditary leadership focuses on maintaining power within a group, women as well as men may have the opportunity to rule, even in highly patriarchal societies. This is obvious in the case of Britain. Queen Elizabeth II is the sovereign of the United Kingdom because the House of Windsor is the ruling family. This characteristic explains paradoxes in the Muslim world related to gender and political power. In Pakistan, for example, the Bhutto family has been an important political dynasty for over two generations. Zulfiqar Ali Bhutto had been Prime Minister until General Zia al-Haqq overthrew him in 1977. Several years later, it was the popularity of the Bhutto family, and the loyalty of its entourage, that led to the election of Benazir Bhutto—a woman—as president of a socially conservative Muslim state. In Benazir's case, kinship reinforced by class considerations carried more weight than gender.[8]

Dictators and Strongmen. Dictators and strongmen obtain power through coercion, and remain in power as long as they are backed by that force, which possesses the means to intimidate and coerce others in the community. As a result, their power can last only as long as they can manage the military and police sectors of a society, while eliminating or coopting any other potential nodes of competition.

A strongman's power does not *originate* from the legitimate investment of authority on the part of the people or organs of state. Therefore, though they might exert great coercive and cooptative efforts to get the institutions of the state to legitimize them, both the durability of their personal legitimacy, and the inter-generational continuity of their rule is always in question.

Hafiz al-Asad of Syria, for example, came to power through a coup-within-a-coup, and was forced to combine authoritarianism with cooptation and self-legitimization to ensure his continued rule, as well as that of his son.[9] Thus, while some might refer to dictators as "kings" or "monarchs" and point to sons succeeding fathers, these methods are different since a royal family has an in-built legitimacy.

As the case of Syria (or Iraq) suggests, dictators and strongmen typically emerge in societies where the traditional leadership and political system has become unstable. Although this form of leadership may fill in a gap for a period of time, it is usually very short lived, lasting, at best, as long as the dictator or strongman's inner circle wishes to remain loyal to his chosen successor. This political instability, for example, brought about the coup by General-turned President Pervez Musharraf in Pakistan, who, like other Pakistani leaders, might prove to be short-lived.[10]

Elected and Selected Leadership. Over thousands of years of world history, the practice of voting for a major formal leader as a condition of him/her taking office is extremely recent. However, selecting a leader on the basis of his or her skills and experience has been practiced for centuries in many economic guilds and organizations. The key feature of elected and selected leaders is that their ability to remain in a position of power is limited to their perceived effectiveness in carrying out their work. This accountability on the part of elected and selected leaders is not typical of many other forms of leadership.

Accountability, in most basic terms, means pleasing the relevant people. Therefore, elected or selected leaders are constrained by popular preferences in their decision-making. This will have implications for Marines and other U.S. authorities, because elected leaders can only accommodate foreign wishes as much as their electors or selectors will allow them.[11]

Culture Operator's Questions: Leadership

The following questions can guide a Marine in understanding the leadership structure in a foreign AO:

- How is decision-making organized, and who makes decisions?

- What are the principles and processes governing policy deliberations and decision-making?

- Whom do leaders have to consult; to whom must they answer?

- How is leadership obtained and passed on (by election, inheritance, demonstration of skill, membership in a certain age or social group, by force)?

- Who are the official formal leaders and what symbols indicate their status?

- To whom do people turn to actually get something done?

- What is the relationship between the formal and informal leader?

Conflicts over Power:
Challenges to Existing Political Structures

For the Marine, who works predominantly in situations of conflict, understanding the cultural organization of power is critical for identifying the source and nature of conflict in a region. By allotting power to certain groups and individuals, all but the most egalitarian political structures marginalize or exclude other groups from the power structure.

The losers in the political structure may accept their lower position for a period of time. However, ultimately in all societies, marginalized members will seek to challenge the authority and position of the leading groups. Although leaders of highly centralized political structures, such as states, may effectively suppress the claims of excluded and oppressed groups for years or even centuries, at some point the locus of power in any political structure shifts. It is during these periods of instability of the ruling group that oppressed and marginalized groups recognize an opportunity to challenge the existing distribution of power. Typically the result is insurgency, conflict and/or warfare.

Many of the conflicts that we, in the U.S., attribute to "tribal," "ethnic" or "religious" differences are in reality conflicts about the relative power of two or more groups within the political structure of a society. In order to understand these conflicts, the Marine needs to understand who the winners are and who the losers are in the current political structure of his AO—even down to the neighborhood level.

As mentioned in the previous chapter, in many regions of the world, power and status are distributed on the basis of social characteristics such as class or kinship or membership in a tribe, an ethnic group, or a religious group. As a result, in many parts of the world certain classes, tribes, or religious and ethnic groups control the society. These privileged groups then enjoy greater wealth, better access to desired goods and career opportunities, and may have the right to control the behavior of lower status classes, tribes, and religious or ethnic groups.

Example: Power, Leadership, and Kinship in Europe

Differential access to goods and opportunities based on social group is not simply an issue in the culturally "foreign" non-West. Historically in Western Europe, for example, power and leadership were concentrated among specific upper class Christian families. Non-Christians and families that did not belong to the aristocracy almost never held positions of authority in the monarchies of the region. As seen in Chapter Five, the French and Russian Revolutions were both challenges to the class and kinship basis of power in Europe at the time (although not to the religious structure).

Both revolutions resulted in restructuring of power, in which kinship and aristocratic class were no longer the primary basis of political authority. In the case of Communist Russia, religion was removed as a form of social and political identity.

Although certain social characteristics can determine one's status and power in society, it is important to remember that most people hold multiple and fluid identities. Remembering the example from Chapter Two of "I.M. Marine," it should not surprise us that people frequently belong to more than one ethnic group or tribe, claim a greater or lesser degree of affiliation with their religious group (or in some cases, may change their religion) and may move between social classes. When a certain ethnic or tribal or religious group suddenly claims its right to power (or challenges the existing power structure), cultural divisions that may normally have been unimportant may become critical. If the conflict becomes extremely divisive, individuals in the society may be forced to select one identity and renounce their membership in other groups. Cultural lines then become drawn and social identities are suddenly rigidified.

European colonial powers often deliberately created rigid tribal or ethnic lines in their colonies as they chose to become allies with one particular group. This polarization of groups had catastrophic outcomes that persisted long after the outside power left.

One of the critical features of conflicts over power by competing social groups is that each group seeks to attract as many members as possible. This explains why a serious genealogy of any tribe will uncover many non-biological descendants of the mythical ancestor. Particularly in times of violent conflict, people living in a particular region may find they are forced to affiliate with the dominant ethnic, tribal, or religious group for protection, regardless of whether or not they have an actual biological or social connection to that group.

Interestingly, as the balance of power shifts, so also will the "immutable" tribal or religious or ethnic identity of individuals in the area. Marines need to keep this fact in mind in their intelligence collection. Far too often we presume that the particular identity claimed by a person in a region is indeed their only identity, and is somehow "real and genetically" based.

We will return to matters of religion, belief, ideology, and identity in much more detail in the next chapter. For now, however, as we consider matters of political association, organization, and power, we need to recognize that in order to swell their ranks, social, religious, and ethnic groups will often call upon an imagined history, making claims to a remembered past and culture to unite their members. Although this rhetoric may seem convincing and is often extremely effective in mobilizing group members, Marines must not be deluded into thinking that the imagined culture and history constitute a permanent reality. It is very easy to explain religious or social conflicts by reducing the issue to: "It's all about religion/ethnicity/tribalism. The Sunni and Shi'i have been fighting since the beginning of time."

The reality is that if one examines the history of the conflicting groups, typically there are periods of time of relative peaceful coexistence, alternating with periods of hostility. At different times in history, for example, Sunni and Shi'i have intermarried, shared political positions of authority, and even worshipped together within certain Sufi brotherhoods. The history that is remembered, then, is the one that supports the political agenda and claims of the group, not necessarily the reality of their past.

As such, Marines are served best by listening to what people in an AO say about themselves and their memberships and identities. But that is not enough, as it might serve as a static smokescreen for evolving realities. Ultimately, then, a focus on resources, groups, power, and inclusion as well as exclusion is just as important.

What does this all mean? Every AO to which a Marine will deploy will possess "have-nots"—or else it would be unlikely that a Marine would be there. This "have/have-not" dynamic is built into virtually all political structures which Marines will encounter.

In such AOs, Marines are usually there because the bare minimum of political, social, and economic enfranchisement has not been granted to the "have-not" population, and the ability of "haves" to enforce their writ and coerce obedience has diminished to the point that "have-nots" have gained some hope of changing the system.

In deploying to these areas, knowing who is included and excluded from political and economic power will be key to understanding, managing, and resolving conflicts. Just as important will be understanding the issues of contention for inclusion and exclusion. Often, understanding these issues—the who and the what—will be much more relevant than ethnicity, religion, or even ideology, which often act as the garb in which concrete group interests are clothed.

Finally, in order to manage or resolve conflict, be it for a week or a year-long deployment, Marines and associated U.S. agencies will need to provide the base-line enfranchisement for "have-nots"—or the purpose for conflict will remain, and U.S. personnel will be seen as part of the problem.

Culture Operator's Questions: Conflicts over Power

For a Marine in a foreign AO the following questions should be asked:

- What are the most important cultural characteristics that determine one's position and power in the community (age, class, gender, tribal identity, ethnicity, religion)?

- What is the degree of polarization in the region with respect to religious/ethnic/tribal identities?

- What is the amount of flexibility and interaction between religious/ethnic/tribal groups?

- Which groups hold power, and to what degree of concentration?

- Which groups are excluded, and along which axes?

- What is their degree of consciousness of exclusion?

- How possible do these groups' leaders think it is to challenge the system?

- How do marginalized and losing groups gain access to valued goods, resources, and opportunities (black market, theft, raids, insurgency)?

- How will alliance with one group affect Marine relationships with the other groups?

Notes

[1] John Burnett, "Violence Surges Along the U.S.-Mexico Border," (Audio Recording) *All Things Considered*, National Public Radio, February 12, 2006: http://www.npr.org/templates/story/story.php?storyId=5203014.

[2] Larry P. Goodson, *Afghanistan's Endless War: State Failure, Regional Politics, and the Rise of the Taliban* (Seattle, WA: University of Washington Press, 2001), 91-133; Ahmed Rashid, *Taliban: Militant Islam, Oil, and fundamentalism in Central Asia* (New Have, CT: Yale University Press, 2001); Robert D. Crews and Amin Tarzi, eds., *The Taliban and the Crisis of Afghanistan* (Cambridge, MA: Harvard University Press, 2008).

[3] For an introduction to the field, see Elman Service and Ronald Cohen, *Origins of the State: The Anthropology of Political Evolution* (Institute for the Study of Human Issues, 1978).

[4] This phenomenon is explored quite well in Lawrence Rosen, *The Culture of Islam: Changing Aspects of Contemporary Muslim Life* (Chicago: University of Chicago Press, 2002). See in particular Chapter Four, "Constructing Political Institutions in a Political Culture of Personalism," 56-74.

[5] Donald Kurtz, *Political Anthropology: Paradigms and Power* (Boulder CO: Westview Press, 2001).

[6] This is in contradistinction to the positivist notion of linear societal development and progress, which underpins much nineteenth- and twentieth-century Western social thought, and American political norms today. See Kenneth Morrison, *Marx, Durkheim, Weber: Formations of Modern Social Thought* (London: Sage Publications, 2006), 26-32, 127-47, 224-44 in particular.

[7] For a discussion of tribes bridging social patterns and political practices, see Lawrence Rosen, *The Culture of Islam: Changing Aspects of Contemporary Muslim Life*, Chapter Three: "What is a Tribe, and Why does it Matter," 39-55.

[8] See Saeed Shafqat, *Civil-Military Relations in Pakistan: From Zulfikar Ali Bhutto to Benazir Bhutto* (Boulder, CO: Westview Press, 1997); and Anne Weaver, *Pakistan: In the Shadow of Jihad and Afghanistan* (New York: Farrar, Straus & Giroux, 2002); Zahid Hussain, *Frontline Pakistan: The Struggle With Militant Islam* (New York: Columbia university Press, 2007).

[9] Volker Perthes, *The Political Economy of Syria under Asad* (London: I. B. Tauris, 1997); Lisa Wedeen, *Ambiguities of Domination: Politics, Rhetoric, and Symbols in Contemporary Syria* (Chicago, Ill.: University Of Chicago Press, 1999); Raymond Hinnebusch, *Syria: Revolution from Above* (London: Routledge, 2002).

[10] Husain Haqqani, *Pakistan: Between Mosque and Military* (Washington D.C.: Carnegie Endowment for International Peace, 2005).

[11] See the now classic treatment of this by Robert Putnam, "Diplomacy and Domestic Politics: The Logic of Two-Level Games," *International Organization* 42:3 (1988).

Chapter 7

Dimension V
Belief Systems

All culture groups have a shared set of beliefs that unite individual members. A **belief** is a certainty, learned through inherited group experiences and practices, about the substance and meaning of phenomena and human activity. An individual's beliefs are relatively immune to influence by personal experiences and the environment. Beliefs work in many ways. As a start, two concern us here. **First**, beliefs influence the way people perceive their world, resulting in a specific world view that structures and affects the way that people in the group interact with each other.

A **second** point is equally important, though perhaps less obvious at first blush. **While it is easily accepted that a group's beliefs cause behavior, the reverse is also true: behavior causes group beliefs.**[1] How does this happen? People create relationships with each other, among groups, and with the environment. These relationships emerge and evolve in the form of structures—social and mental. Subsequently, in a group of people these relationships and structures need to be explained or justified as normal and proper. Cultural beliefs perform this function. Finally, these beliefs condition the way later generations of the group understand the substance and meaning of phenomena and human activity as related to their particular relationships and structures.

This second point is quite significant. As the environment, behavior, and social structure of a culture change, cultural beliefs will also shift to support the new reality. For example, in the 1920s in the U.S., cultural beliefs dictated that the appropriate role for women was wife and mother, and that work outside of the home should be done by men. During World War II, however, as men left the U.S. to fight in the war, women began taking the place of men in factories, businesses and even state-side military activities. The new social reality of women in the workforce (along with other social and economic changes at the time) forced Americans to adjust their cultural beliefs; and over the next sixty years, many Americans adopted a new attitude, which was more tolerant of women working outside the home.

However, this new group belief about the role of women did not occur overnight—it has taken generations. This is because *individuals'* retention of beliefs retards the pace of change in *group* beliefs. Thus, in an area of operations, Marines should never assume that changing realities on the ground are reflected in meaningful change to beliefs *during their tour*. People with whom Marines interact will cling to beliefs that seem out of sync with realities that Marines and they themselves created. Put differently, Marines will find that indigenous people will preserve beliefs that appear out of step with their own evolving behaviors.

When a Marine enters a new AO, he or she is unlikely to understand the peoples' beliefs quickly in the absence of significant, region-focused study and experience. However, cultural beliefs and ideals are ultimately expressed at least partially through observable words and actions: through historical stories and myths; symbols, rituals and icons; and taboos governing acceptable or forbidden behaviors.

Americans often associate beliefs and ideals with formal religious systems as reflected in religious writings, ceremonies, and traditions. However, both in the West and many other parts of the world, informal religious systems, practices, and groupings illuminate a community's beliefs and ideals. Even more important, the Marine must remember that **neither organized religions nor informal religious practices are the only producers of beliefs and ideals**, nor are they the only place where beliefs or ideals are found.

Furthermore, though all culture groups have shared beliefs uniting members, Marines interacting with real live members of a group in a foreign environment should remember that not everyone in that group has all of the same beliefs. As we emphasized in Part I of

Operational Culture in particular, people, though members of culture groups, have freedom of choice and belief. Therefore, by approaching cultural beliefs from a more pragmatic position of evaluating observable behaviors and writings, rather than trying to "get into the mind of the Arab or Muslim," for example, Marines can begin to understand the day to day local cultural ideals where they work, and the variability of beliefs and attitudes among the actual people there.

Some Features of Beliefs and Symbols

History, Imagined Memory, and Folklore. Most basically, **history** is what happened in the past. The history of an AO is important to Marines. However, in an operational culture sense, history is not important because of what happened in the past. Rather, it is how people selectively remember their past, and what meanings they choose to derive from that past, that are of primary concern to the Marine. The selective remembering of events, and belief-influenced attribution of meaning to them, is known as **imagined memory**. In a group context, imagined memory can be called **collective memory**.[2]

Historical stories (and quasi-historical tales such as myths, legends and folklore) are important keys to revealing underlying cultural themes and beliefs. In simple terms, **folklore** is a group's collection of stories, sayings, and narratives of history passed down through the generations. Each generation receives this inheritance, imbues it with new meaning, and adds new narratives based on new collective experiences.

Several ways of evaluating the cultural meaning of history are useful for Marines to consider:

Memory as Constructed. Memories come into being because someone sees an historical event and then gives it meaning and importance. Indeed, events can be remembered even though they did not happen, because they serve a psychic purpose.

For example, a famous story from the Spanish-American War about the Battle of San Juan Hill describes Theodore Roosevelt charging up San Juan Hill on a horse. Historians today point out that Roosevelt and his "Rough Riders" actually ran (not rode) up the nearby Kettle Hill—and some suggest that Roosevelt never even made it to the top. The former memory, however, serves the psychic need for an image matching the late nineteenth-century American self perception of being a nation epitomized by the physically tough "every man" who could tame nature and beast while conquering foreign (and foreign-sounding) realms, in effect extending manifest destiny beyond the shores of the Pacific Ocean.

Memory as Ideological Fabrication. Sometimes, outside authorities build stories around real or fake events, in order to fabricate a memory that supports their agendas. These often involve ideas of continuous existence on a patch of land; ethnic or biological resemblances among all people of the "true" nation; or the celebration of a significant historical personality or event that embodies all desired traits among members of a community.

For example, the Italian Fascist regime of Mussolini chose as its symbol the ancient Roman Fasces, attempting to associate the 1920s-40s Italy with the Ancient Roman Empire, and Mussolini with the Roman Emperors. Likewise, the cry "Remember the *Maine!*" helped foster support for the 1898 Spanish-American War. The saying implied Spanish treachery as the reason for the explosion of the USS *Maine*. In fact, an accident resulted in a coal bunker explosion.

The Event Evokes the Memory and the Meaning. Since historical events have more than one part to them, the same event can have more than one meaning for people, as they choose to focus on individual components and attribute particular meanings to them. In thinking of the importance of history as a reservoir for memory creation, therefore, it is important for the Marine not to assume a single meaning behind each story.

Likewise, memories are activated in the here and now based upon which particular event confronts a person. There is no particular history-driven logic determining what people choose to call upon in their past in the course of daily life. Rather, the logic of memory is almost wholly related to the present tense—challenges, opportunities, and interpretive needs that individuals and communities face today.

In the same fashion, what people consider useful about a past event, and how they choose to remember it—the lessons they find in the past—is all about what they seek in the present tense. In an operational culture sense, "history" is a tool at the disposal of actors in the present. Marines therefore must learn history, not for its own sake, but in order to understand what groups within an AO might use that history; potential values placed on that history; and circumstances in which those meanings and uses can be put into action for present-oriented goals.[3]

Example: Events and People Make the Memory's Meaning

On the tenth day of the Arab-Islamic month of Muharram in 680 C.E., Hussein bin 'Ali traveled with a reinforced platoon-sized band of warriors from Medina eastwards to Kufa in Iraq, to raise an army against political opponents in Damascus. On the way to Kufa, he was met by the entire army of the Umayyad dynasty. In a lopsided battle at Karbala, Hussein was defeated and his entire army massacred.

Hussein was the son of 'Ali, who in turn was the son-in-law of the Prophet Muhammad. Upon the death of Muhammad, some Muslims felt that 'Ali, as a relative of the Prophet, should become the successor (caliph) to lead the Muslim community. 'Ali, however, was passed over and 'Umar, a close companion of Muhammad, was selected instead. Those who supported 'Ali formed a faction (shi'a), arguing that righteous leadership of the Muslim community should be based on descent from the Prophet's family. Part of this faction lived in Kufa in Iraq. They invited 'Ali to Kufa, and promised to raise an army to restore the rightful family line to power in Islam. However, the army never formed, and it is thought that Kufans had lost nerve, and were afraid to rise up against their opponents.

"Karbala," therefore, has many meanings, depending on who remembers it, and for what purposes. At different times, Lebanese, Iranian, and Iraqi Shi'ites have remembered Karbala as a time when a righteous Muslim—Hussein—stood up for justice and the straight path, even against obviously insurmountable odds.

Example: Events and People Make the Memory's Meaning, continued.

Particularly in Iran and Iraq, Shi'i leaders could tap into this memory of Hussein and Karbala to rally their flocks against the secular Shah or Ba'thist Saddam, both of whom possessed a monopoly of brutal force.

A different association with Karbala exists as well. It is the more traditional one associated with the holiday commemorating the event, 'Ashura (tenth of Muharram). Here, Shi'is remember the cruelty of the material world of the present tense, emphasizing the need for loyal piety while awaiting the return of Islam's messianic figure. Furthermore, ritual self-flagellation, and retelling the entire Karbala story, highlight the infidelity of weak-hearted Kufans who failed to stand up for what is right, and did not come to Hussein's aid. This, in effect makes all Shi'ites complicit in the Kufans' crime, but is also a challenge to Shi'is today to do better than their forebears.

There are, then, multiple ways to remember Karbala. The key point, however, is that the conditions of the present tense often color how it is remembered. Why, for example, remember Karbala in the absence of violence and political exclusion? Likewise, whether an Iranian Shi'ite is inspired by the Hussein martyrdom part or the Kufan infidelity part of Karbala is determined by what challenges they face on any given day.

It is thus no coincidence that the relevance people find in Karbala has grown in the past fifty years. This awareness is not culturally "embedded" or "programmed;" but rather results from evolving political agendas and social trends. In Lebanon, for example, Shi'ite political maturation progressed in the midst of political exclusion during the 1970s–80s. Since the 1980s, Saudi Arabia, Bahrain, and Qatar have seen growing awareness on the part of minority Shi'ite populations of their own disenfranchisement *and* abilities to destabilize unyielding regimes.

The ideological invigoration as well as politicization—and now armed power—of Iraq's Shi'ites could only occur after the post-1991 weakening of the Ba'thi regime in Baghdad. Just as important in that evolution was the Iranian revolution of 1979. At that time, the Ayatollah Khomeini had called upon Iranians to rise up like Hussein, and their positive response was possible because secular Iranian intellectuals had re-popularized Hussein-Karbala as an appeal for equality and political awakening (1960–70s).

After 2003, internal Iraqi changes came together with ideological changes since Khomeini's call to empower the Shi'ites of Iraq—who could now assume the role of the collective "Hussein," in order to claim their rightful place against both the minority Sunni elites and the foreign (Christian) occupier.[4]

Like history, folklore often reveals underlying cultural values and ideals. These ideals are passed down from generation to generation through oral as well as written traditions: in folktales, legends, and popular sayings and expressions. While the analysis of the folklore of a cultural group can take a lifetime, Marines can easily pick up some of the more important local cultural values just by listening to commonly repeated sayings, expressions and stories.

In the U.S. for example, the Puritan notion that we can control our lives and experiences through hard work is repeatedly emphasized through such common sayings as:

"Early to bed, early to rise, makes a man healthy, wealthy and wise."

"God helps those who help themselves."

These sayings contrast significantly with popular expressions in the Middle East, which emphasize submission to God's will and acceptance of a divine plan that one cannot control:

Insha'allah
> "If God wills it." (Arabic, Turkish, Persian)

Im yir-tzei HaShem
> "If God wills it." (Hebrew)

Agar reza-ye Khuda bashet
> "If it finds the favor of God." (Persian)

Keh deh khoday reza wi
> "God-willing." (Pashto)

Allah A'lam
> "God knows [the outcome—not I]." (Arabic)

Allah ma'a al-sabirin
> "God is with the patient." (Arabic)

Significant for Marines, particularly those working in Information Operations, is the way that popular folktales, sayings, and imagined histories are used in propaganda to rouse popular sentiment. The Marine contemplating an IO campaign based on folklore, however, would be well-served to remember that folktales and memories possess different meanings and significances for different communities within an AO. We will return to this shortly in our discussion of icons.

When operating in a specific area, it is important for Marines to grasp its history—not for what happened, but for how people think of it: what the collective memories are and how they color attitudes to Marine actions. Similarly, Marines need to track the way that folktales, legends and sayings are used in an AO to reflect local perceptions of Marine behavior in the region.

Culture Operator's Questions: History, Imagined Memory, Folklore

• Several questions that Marines should ask when working in an unfamiliar region are:

• What are the pivotal historical stories that all people in the community share?

• How do different groups in the AO give different significance to the same historical stories?

• What are the daily sayings and folktales that everyone refers to in common conversation?

• How are these remembered histories, folktales, and sayings used to emphasize or teach important values and ideals?

• How are these histories, folktales, and sayings used to support propaganda for or against Marine and U.S. activities in the region?

Icons. Prominent in both folktales and remembered histories are icons: individuals who become larger than life and symbolize many of the positive (or negative) values of a society. Historically, icons were physical objects felt to represent a deity or holy person. In the Orthodox Christianity of Byzantium (and later Greece and Russia), *ikonos* were objects with representations of the Virgin Mary, Jesus Christ, or other saints, through which people channeled prayer.

Today, icons are often heroes with almost superhuman abilities—the mere mention of their name can evoke a shared memory and inspire certain kinds of actions or attitudes among a group of people. The icons thus become role models, helping members of a cultural group to learn—often without being aware of it—the values of that group.

While we will consider the Marine Corps as a culture group in the Conclusion, a perfect example of an icon in the Marine sense is Gunnery Sergeant Basilone—"Manila John"—who during WWII received the Medal of Honor for heroic and selfless actions in killing the enemy and saving Marine lives, ultimately giving his own life in his last efforts, winning the Navy Cross.[5] A recent example similar to Basilone is Corporal Jason Dunham, awarded the Medal of Honor for diving on a grenade in western Iraq April 2004 in order to save his fellows. Both are held up as exemplifying all it means to be a Marine: mission oriented to the last, dismissive of danger, and committed to their fellow Marines to the point of the ultimate sacrifice. Icons like Basilone and Dunham are role models, showing recruits and young Marines what it means to be "of" the Marine Corps.

Even more so, there are phrases and maxims associated with particular Marine icons, including Lloyd Williams, "Chesty" Puller, "Howlin Mad" Smith, and Dan Daly, for example. These maxims furnish core components of the creed of what it means to be a Marine:

"Retreat? Hell, we just got here." (Williams)

"We're surrounded. That simplifies the problem." (Puller)

"They are in front of us, behind us, and we are flanked on both sides... they can't get away from us now." (Puller or Lewis)

"Come on, you sons of bitches! Do you want to live forever?" (Daly)

Such icons become not only larger than life, but longer-lived than their lives: Marines recite Lejeune's birthday message every anniversary, and are known to say, "Goodnight Chesty Puller, wherever you are."

Icons are not, however, single-valued—just like historic events, or stories, have different meanings depending on who is doing the reading, or depending on the current circumstances. A single icon can be a villain, hero, or role-model for certain values, all at the same time. This is because icons, like remembered history, ultimately tell us more about the identity and the values of the culture that remembers it than the facts narrated.

Example: Multi-Valent Icons

Ghengis Khan led Mongol hordes out of Inner Asia in the beginning of the 1200s, wreaking havoc through Central Asia, Iran, the Fertile Crescent, the Levant, and into Eurasia and Central Europe. For Central and Eastern Europeans in particular, even naming Ghengis Khan continues to evoke horror, and like "Huns," can be used to tar opponents. Both names are associated with barbaric torture and devastating destruction of peoples and cities they conquered as they moved Westward from Mongolia. In European movies and folklore, Ghengis Khan epitomizes a ruthless, merciless, greedy, uncivilized man who is the antithesis of genteel European civilization, stooping to almost animal-like behavior in his gratuitous violence and debauchery, treatment of his enemies, and his disregard for the art and architecture of the cities he demolished.

Ghengis, however has a second history recounted by the Mongolians who see him as a source of national pride. For the Mongolians, Ghengis Khan is the epitome of a brilliant, strong, fierce, and strategic leader. Songs sung today around hearths in Mongolian tents recount his brave deeds and legendary conquests. For a Mongolia long colonized by Chinese, Russians, and Soviets, Ghengis—who also emerged from a land dominated by others—is an icon for physical/national strength, and worthiness of independence as well as respect for past accomplishments.

Example: Multi-Valent Icons; continued.

There is also a third way to remember Ghengis—in this case Cengiz Han. In contemporary Turkish official memory, Cengiz's onslaught was a "Mongol-Turk" empire-making effort. Cengiz and his "Mongol-Turk" tribal warlords built a state that restored stability to the Middle East, resuscitated regional economic exchange, and developed the bureaucratic, political, military, and intellectual superstructures of how states needed to work in Eurasia. As such, the Ottoman Empire, and perhaps even the Turkish Republic, are now using some of those very state-making tools of Cengiz Han, the "Mongol-Turk head of state." It is in the comparison of these three views that we find the contrasting values of the three societies, as well as the near-similarities of the latter two.

In entering a new Area of Operations, Marines can learn much about local beliefs by determining who are the important icons—historical figures, mythical figures, and present-day heroes.

Culture Operator's Questions: Icons

The following questions can be helpful, particularly for those working in Information Operations:

- Who are the local heroes? What important qualities do these heroes embody?

- Who are the local villains? Why are they villainous (what makes them evil)?

- Are the heroes or villains compared to Marines or Americans?

- What do the comparisons illustrate about local attitudes towards the U.S. and the military?

Symbols and Communication. One of the unique characteristics of humans is their ability to communicate using symbols. Symbols can be physical, such as a flag; verbal, as in a spoken language (which is a symbolic string of sounds); or behavioral, as in non-verbal communication such as smiling. The unique feature of symbols is that they have meanings that are culturally dependent. The word "Achoo" in English, for example, symbolizes a sneeze. In Latvian it means "Thank you." To symbolize purity, an American woman wears a white wedding dress; her Chinese counterpart wears a red gown. Whether physical, verbal, or non-verbal, symbols communicate essential information about the beliefs and identities of local people. In order to understand and work with local communities, Marines need to understand the basic symbols around them.

Physical Symbols. Physical symbols include any physical objects that hold a symbolic meaning greater than their practical utility for the people in a group. Frequently, physical symbols, both natural and man-made, mark out status and/or identity. Clothing, material adornment, and symbolic objects may all indicate one's social membership and identity, as do physiognomic elements such as hair, piercing, or scars. In Sub-Saharan Africa, for example, facial scars on adult males distinguish Gorani tribes from Fulani, or Tuareg, and are a matter of status. Stars on a Marine's collar also communicate status—as a General Officer.

Not only do symbols provide information about status and roles to members within a group, but they also indicate who is in and who is out of a group. Within the Marine Corps, for example, a second lieutenant reads a gold oak leaf in order to know who is a major, and how s/he is to act toward that major. Outside the U.S. Marine Corps, however, a French major may not be able to read the gold oak leaf of a U.S. Marine major—he and the Marine thus know they possess different group membership. As such, physical symbols may also serve as visual indicators of certain aspects of social structure. For example, an Afghan male who displays a ball-point pen in the breast pocket of his *shalwar kameez* may be indicating his status as a member of the literate, educated elite, with certain expectations of deference and treatment. We discussed these elements of social structure in Chapter Five.

Just as a new Marine quickly learns to read the meaning of symbols on his commanding officer's collar, Marines in foreign AOs need to be able to recognize the physical symbols that distinguish individuals within the community in order to operate effectively.

As we alluded to in our discussion of the environment in Chapter Three, physical symbols can also serve to mark space symbolically: indicating ownership or membership, or a religiously important place. Heaps of stones may symbolize a sacred spot in Tibet. A flag planted on the moon symbolizes U.S. presence there. A fence symbolizes a boundary which one should not cross. In many cases, the "symbology" might not appear in images to which Marines are attuned: the pile of rocks might look more like unarranged refuse; a symbol of local sovereignty might look more like tangled twigs. Still, it is essential to remain alert for indications of physical symbols, to develop a more nuanced sight picture of the environment, reflecting the views of indigenous people.

Perhaps one of the most extraordinary sets of physical symbols created by man is written language. While an outsider may only see a bizarre set of twisted lines, to someone in China the many characters convey important meanings. Written symbols are everywhere in our modern world. Signs on stores, political ads, social announcements, and even graffiti all tell the reader important information. Indeed, for a Marine operating in China, some symbols such as the ones representing male and female on the doors of bathrooms are definitely worth learning.

While characters themselves are a form of communication through symbols, that is only one aspect of the symbology of the written word. For example, red ink in the margins of a written text symbolizes something immediately quite understood by western readers—it is critique and/or correction. Likewise, a note jotted in pencil on a piece of photo-copy paper has a very different significance from something written in ink on very heavy card stock paper embossed with the author's initials.

Again, even though Persian uses the Arabic alphabet, a foreign speaker of Arabic or a semi-literate Arab will immediately recognize the particular Arabic *font* normally used in *Persian* as symbolizing something linguistically and culturally foreign. Thus, when coalition forces in Iraq inadvertently used the *Persian* font in their *Arabic*-language IO flyers, they did not achieve their intended effect on the anti-Iranian local Sunni population.[6]

Verbal Symbols. Most of us take language for granted. But speech is an extraordinarily complex interchange of sounds that hold symbolic meaning. As any Marine who has spent time in a foreign region knows, the ability to communicate in the local language opens up an entire world of understanding that is closed to foreigners who do not understand the language. Though foreign language fluency need not be a skills set resident within all individual Marines or Marine units, it is important to operational success to consider what foreign languages are spoken in what parts of the AO. Further, there are certain baseline language skills required in a unit to communicate needed messages to indigenous partners or neutrals. Likewise, within one's unit, or an adjacent unit, there are often heritage language speakers.

Many languages feature both the standard language and several dialects, spoken in different parts of a country; they can indicate regionalism, rich vs. poor, schooled vs. uneducated, male vs. female. Understanding these differences, and knowing how to look for them, will better help Marines to know who their interlocutors are, who belongs, and who is a "stranger." For example, though a Semitic language, the Aramaic of Christian Baghdad sounds different from Iraqi Arabic. Without even knowing Iraqi Arabic, Marines have sometimes distinguished Iraqi Christians from other ethnic and social groups based on the sound and cadence of the language they hear.

Furthermore, in a cultural group there may be sub-groups who use sub-group-specific terminology or jargon—as is clear from the U.S. Army's "huwa" or the U.S. Marine Corps' "oo-rah." In fact, these "non-standard" usages or terms can tell the Marine much about the background, affiliations, or aspirations of the indigenous person with whom he or she is interacting.[7]

Example: Language as Symbolic Signaling

In parts of North Africa, native speakers of Arabic make a conscious effort to speak French with authorities or foreigners, in order to indicate their education and credibility. Even if that person's knowledge of French is clearly defective, he/she will use it whenever possible, dropping words of it into their Arabic, or switching back into it if the other speaker uses Arabic—even if that fluent Arabic speaker is a foreigner who knows no French.

In these cases, it is not the subject of the communication that drives use of French; rather, the North African is signaling status, outlook, and, to an extent, group membership. If a Marine knows this, he might be able to accord the North African the status and treatment that would make him positively inclined to the Marine.

This would be particularly true for Marine interactions with Tunisian (or Algerian) military officers, for example. While one might be inclined to study Tunisian Arabic before an exchange visit with the Tunisian military, by using that dialect with a Tunisian officer, a Marine might in fact communicate what would appear to be an assumption that the Tunisian was uneducated or "uncultured." The Marine might also communicate that he himself was uneducated, and unworthy of the Tunisian's time. So in this case, learning some French, and using it with military counterparts from an Arabic-speaking country, would be an appropriate verbal symbol of the Marine's status, education, and the positive value he/she puts on working with the Tunisian military.

Non-Verbal Communication and Symbols. In addition to the physical and verbal symbols of communication, people around the world have developed unique non-verbal, behavioral symbols that convey important meaning in social contexts. Some of these non-verbal symbols such as smiling and laughing and crying appear to be shared across virtually all societies around the world. Other non-verbal symbols, however, vary from culture group to culture group. Indeed, some symbols have very different meanings in different cultural contexts. For example, the practice of forming an 'O' by joining the thumb and index finger is a positive sign indicating 'OK' in American society. In some parts of Asia, however, it is an offensive sexual symbol.

Example: Non-Verbal Symbolic Communication

Among some Russian communities, clasping one's hands and shaking them over each shoulder a few times simulates hugging a good friend. It is thus a symbol of amicability, though when Khrushchev did it in front of the U.S. press, it was seen as a warlike, provocative move.

As a second case, in certain European circles, attending a political event with a large black umbrella is an accusation that the political leader is surrendering to the opposition. This relates back to Neville Chamberlain, British Prime Minister at the 1936 Munich Summit with Nazi Germany, who appeased Hitler. Emerging from the summit with a black umbrella, he declared "Peace in our Time." In fact, when Israeli diplomats came to the United States after signing the 1993 Oslo Accords with the PLO, Americans who opposed the Accords would open black umbrellas during the Israeli leaders' appearances.

In this case, most Americans had no idea what the black umbrella meant; the Jewish Israelis, however, with a memory of the Holocaust, fully grasped the political point.[8]

Non-verbal symbolic communication is often described in terms of body language such as hand and facial gestures. However, spatial dynamics can also be a central part of the non-verbal component of communication: How people place themselves and where can often create the context or provide substance of communication. Seating patterns, the order or arrival of guests, the sequence of greeting guests, who is served first and who does the serving—all of these establish, reflect, and reinforce social hierarchies. Non-verbal communication therefore plays an important role in meetings and social interactions such as negotiations, celebrations, funerals.[9]

Non-verbal behavior is also significant in communicating ideals such as bravery, fear, or honor. For example, The U.S. military tactic of firing from cover, considered operationally prudent, is considered in some areas, such as Chad and Afghanistan, as cowardly. Put differently, actions communicate—intent, desire, attitude, and, indeed, ideals and values. Not only is it incumbent upon a Marine to learn about these from AO to AO, but it is also essential to understand the *implications of Marine actions*—from how one sits to how one arranges vehicles—*as communication*. Most fundamentally, the Marine must remember, and train his or her Marines to remember, that **every action is IO.**

However, because cultures are not homogeneous; because communication is human and therefore imperfect; and because at different times different people mean different things through their body language and non-verbal communication, even within the same culture group, people can misinterpret symbolic communication involving body language. For this reason, while "do-this-don't-do-that" approaches to extremely visible, "performable skills" might be given to check lists in a military training context, it is important to realize that symbolic communication through body language is not an exact science: One does not get it "right," and it ought not to be a central issue of planning and operating, compared to the larger, more conceptual issues under discussion in this and other chapters.

Example: The Checklist Approach to Non-Verbal Communication

During his pre-deployment training, a Marine commander had been instructed not to wear sunglasses when talking to interlocutors of his social rank in Iraq. One day in Iraq, while meeting a police chief in the street, the Marine commander took off his sunglasses to greet him. The Iraqi, who left his own sunglasses on, responded by saying, "It's hot and the sun's glaring. Why would you take off your sunglasses?" This same commander had been instructed not to use a very firm American grip when shaking hands. When he greeted an Iraqi civil leader with a soft handshake, the Marine in fact made an impression of physical weakness and boredom.

Likewise, when an Iraqi officer clapped him on the back with his left hand, the commander assumed this meant the Iraqi disliked him, because he had been told that the left hand signified contempt in "Islamic" culture. Contempt was, however, the opposite of what the Iraqi had intended.[10]

The micro-level fixation with observable "do's and don'ts" that can be catalogued as performance tasks might obscure the larger implications of symbolic communication. In many cases, commanders have missed symbolic communication's more fundamental importance, as they focus on minute details of a specific behavior assumed to be culturally "loaded." We will return to this discussion in the context of training and education in Chapter Nine.

Culture Operator's Questions: Symbols and Communication

In order to successfully communicate and negotiate with local people in a foreign AO, Marines should consider the following questions:

- What physical symbols (clothing, headdress, scarification, insignia) indicate membership or status in the ethnic, religious, and social groups of the region?

- What physical and written symbols (signs, graffiti, fences, spiritual markers) are important to be able to recognize in order to navigate and understand what is happening in the region?

- What words or phrases are essential for basic communication with local people?

- What non-verbal behaviors may be misinterpreted by local people? Which non-verbal behaviors are important to understand in meetings and negotiations?

Rituals. Rituals also offer the Marine a window into cultural values and ideals. Rituals are often characterized by the notion that the actions in the ritual themselves must be performed a special way to be valid. For example, during a change of command the honor guard, reading of command authorization, and passing of flag of command—each of which is a single, specific ritual—all have to be performed properly for the change of command to be considered valid.

Also of note, in religious rituals, the necessity to perform every step of every ritual correctly is related to the conviction that the spiritual soundness—or metaphysical validity—will only be preserved if each step of that ritual is performed correctly. In these cases, not only is imperfect performance shoddy, but it also violates holiness itself.

For a Marine, rituals and ceremonies are not just quaint local traditions but important for three key reasons:

- Knowledge of ritual performances in the AO will allow commanders to ensure proper measures are taken by Marines to show respect and understanding towards indigenous people.

- The enactment of rituals reflects and reaffirms the values and identity both of the person performing the ritual and of the community of which he or she is a part. Therefore, Marines can observe rituals to know, according to local people, "who we think we are" and "who is not a part of us." This is all about group membership.

- Knowledge of rituals is important, because of their sociopolitical context and meaning to those involved—and to those observing from outside the community.

Example: Ritual as Social and Political Communication

Earlier in this chapter, we examined the culturally important meaning of the Karbala/'Ashura story for Shi'ites in Iraq and beyond. Understanding the importance of the annual 'Ashura ritual-enactment itself is equally important for Marines. Every year at 'Ashura, tens of thousands of Shi'ite pilgrims descend on Karbala.

They re-enact the tragedy of Hussein through passion plays, with some participants engaging in acts of self-flagellation. First, knowledge of these rituals permits commanders to prepare for the pilgrimage and deploy forces in the proper fashion so as not to inhibit ritual performance while providing sufficient security and force protection.

Second, as we saw earlier in this chapter, on a yearly basis, 'Ashura re-commits Shi'ites to their self-identity as persecuted martyrs for the cause of righteousness. It would certainly assist a commander to understand that the self-perception of a group of Shi'ites he is addressing during the 'Ashura season is informed by a sense of persecution and dispossession by outsiders, and that certain rituals reaffirm this self-perception.

Example: Ritual as Social and Political Communication, continued.

Finally, the commander and ground-level Marine need to grasp that the mere performance of the 'Ashura rites is in itself a political act and statement, both within the Shi'ite community—and to those outside the Iraqi Shi'ite community, meaning Iraqi Sunnis. Several times in Iraq's history, most recently under Saddam, 'Ashura was forbidden by the (wealthier, more politically connected) Sunni minority.

One could therefore argue that the externally-directed communication of the ritual is even more important, to the effect that "we Shi'ites are here, we are taking what is our due... in religious, social, and political terms." This knowledge helps Marines to see the 'Ashura ritual as a statement of Shi'ite communal-political prominence—a statement that is contested by Sunni Iraqis.[11]

Culture Operator's Questions: Rituals

When observing a local ritual or ceremony, a Marine can learn about local cultural values by asking the following questions:

• What behaviors and actions are important in the ritual or ceremony, and what does this reveal about cultural ideals and values?

• Who participates in the ritual, and what roles do the participants play?

• What does presence of participants, or the nature of their participation, say about their membership and status in the group?

• What does the public performance of the ritual communicate to outsiders?

• How is this performance potentially a politically charged statement about the group's status and rights within the larger society?

• What activities, not related to the ritual or ceremony itself, occur at ceremonial gatherings, due to the social status of the participants?

Norms, Mores, and Taboos. All cultural groups have written or implicit rules about what is acceptable and unacceptable behavior. Frequently, in military cultural training seminars, these norms are emphasized, often to the exclusion of many other aspects of Operational Culture. Understanding the do's and don'ts in a new region can be important in creating goodwill with the local peoples. And in this text, we encourage Marines to learn the proper social etiquette in an area. However, we caution Marines to recognize that a few courses on the correct way to serve tea in Afghanistan are unlikely to give leaders the full spectrum of cultural knowledge necessary to develop a successful operation. Despite this caveat, successful Marines should have a basic cultural understanding of an AO's social norms, mores, and taboos.

Social Norms. Social norms are cultural expectations about how one ought to behave in a given situation. Norms can be seen as social guidelines that most people usually follow. However, norms are not rigid, and people may accidentally or deliberately ignore the norms with only minor repercussions (such as disapproving looks, avoidance by others).

In the U.S. for example, Americans try to adhere to the norm of punctuality for meetings. The accepted norm is that people will arrive at a meeting at a specified time, and use that specified time—and only that time—for that specified purpose. However, while the U.S. *military* adheres fairly rigidly to this norm, it is understood that for *personal* meetings people may show up late, or sometimes not at all. Although others may complain or criticize someone for being late, except in extreme cases, people are not severely punished for their tardiness. Americans do not jail or kill people for failing to adhere to the norm of punctuality. In short, a norm is the "proper thing to do," but disregarding a norm is not a fundamental social transgression.

Norms are often an obvious point of failure in inter-cultural communication—after the fact. For example the Western belief in individual efforts and meritocracy can conflict with ideas of collective, kin-based solidarity, where standing apart from the tribe contravenes the local code of honor. In Zimbabwe, the U.S. Embassy "Foreign Service National of the Year" award was quite unpopular among indigenous people, who felt singled out, contrary to a creed of group loyalties. After more experience in the region, U.S. Foreign Service Officers have moved away from this practice, realizing it is not in accord with local norms.

Yet even within the same region or culture, attitudes about acceptable behavior and values are often variable. In the U.S., "bravery," "justice," and "security" are examples of values that are understood differently based on people's different "standards"—or interpretations based on past experiences and constraints. For example, understandings of acceptable levels of "law and order" in southern California cities are very different from those in towns along the Texas-Mexico border, given different experiences and capabilities.

Mores. A more (pronounced *mor-ay*) is an implicit or explicit rule regarding permissible or forbidden behavior. In contrast to norms, violations of a cultural more usually provoke serious repercussions. Social ostracism, physical attacks, or even death may result from challenging mores. Cultural mores vary over time and place, and Marines should not assume that behaviors and ideals that are accepted in the U.S. will be regarded in similar ways around the world.

As an example of a more with serious cultural and physical repercussions, in Mediterranean countries a high value is placed on the virginity of an unmarried woman. Historically in Italy, Spain and Greece, and today in a number of the countries of North Africa and the Levant, an unmarried woman who gets pregnant is considered a disgrace to her family. The punishment for this transgression (men are rarely punished) can range from the girl being cast out from her family to death. Clearly this is not a norm that can be disregarded as "bad form;" rather, it is a more whose violation bears severe consequences in a particular culture group. This is in contrast to American culture, where premarital virginity is considered a norm, but not necessarily a more.

Mores may include codes of honor—though codes of honor may also communicate norms. These codes often provide a basis for assumptions about how things should/do work in an area, just as they influence people's judgments of events. Though Afghan Pashtuns are Muslim, for example, it is often the Pashtunwali, an unwritten code of behavior based on an exaggerated sense of personal honor, which trumps religion in interpreting events and determining proper actions in the region. As a code of honor, Pashtunwali includes honor (*nang*), revenge (*badal*) for injury to honor; and asylum (*nanawatey*), or granting of hospitality to those who seek refuge. Pashtunwali is therefore an unwritten, culturally-coded list of mores.[12]

Taboos. Taboos are activities or the use of physical objects that are explicitly forbidden. Taboos are generally based on religious notions of permissible and impermissible objects and activities. In contrast to mores, taboos are rarely about "what you should/must do," but are about "what you should/must *not* do." In a seeming paradox, however, broken taboos may not always carry the heavy repercussions of violations of a social more. This may, in part, be attributed to the notion that God or the powers that be will strike down the offender, so that others in the society need not enforce the requirement of observance.

In this sense, the factor of public restraints vs. private freedom is at play: with respect to taboos, it is often the *public* violation of them that causes the particular disgust of society. In more secular Muslim societies, significant numbers of Muslims choose not to fast during the month of Ramadan. Respecting the norms and taboos of Islam, however, they ensure that their personal choice does not become a publicly visible practice. In Tunisia, for example, it is widely known that some people do not fast during Ramadan. However, *public* violation of the fast elicits strong displeasure and sometimes social ostracism. For example, even though he was fabulously popular as the hero of North African nationalism, Tunisian

President Habib Bourguiba's drinking of orange juice on *state television* still received considerable public censure—or praise as an iconoclast—in Tunisia and beyond.[13]

In addition to the public-private difference with respect to taboos, there are also different levels of taboo. Clearly, drinking orange juice during Ramadan cannot be compared to the taboo of incest—which, though defined differently in different cultures, is never considered acceptable "behind closed doors," and is legally and/or physically punishable in most cultures.

Probably the most obvious taboos that a Marine will encounter are taboos regarding food and drink. Many cultural and religious groups around the world have rules regarding forbidden foods. Muslims forbid drinking alcohol and eating pork (plus a host of other "unclean" animals) and require that animals are killed according to certain *halal* rules. Likewise Jews, Hindus, Buddhists and certain Christian groups have their own dietary restrictions.

Other equally important taboos center around notions of physical purity and cleanliness. Many cultural groups have taboos forbidding "unclean" individuals from participating in certain social activities. Traditionally in India, the Untouchable caste was not permitted to socialize or interact with other castes due to their unclean work as garbage cleaners. And in parts of East and West Africa, menstruating women cannot prepare food or participate in the community's social life due to notions of their impurity.

This discussion of norms, mores, and taboos has not touched upon laws or the legal system of a country. This is because culture groups in AOs to which Marines will deploy may not have formal legal systems or codified laws unto themselves—though they might have sub-national norms, mores, and taboos that are unique to them. Legal systems are a product of a state society, and—in the model most recognizable to westerners—are imposed upon all residents of the state, regardless of the sub-national culture group's practices and traditions. However, since state legal systems do not account for all local norms, often local culture groups disregard laws that do not reflect their beliefs, or develop "understood" laws to enforce their mores and taboos. Conversely, there are cases where the state itself recognizes the validity of—or must accept—alternative local legal norms, practices, or structures.

Example: Legal Systems within Legal Systems – The Native American Case

Many governments have struggled with how to apply laws to indigenous cultural groups, whose beliefs and traditions are different from those of the people governing the state. One solution has been to allow indigenous groups (particularly differing religious groups) the right to apply their own laws in matters not affecting the state. Like other pre-modern dynastic states, the Sunni Muslim Ottoman Empire, for example, governed according to the *millet* system, which granted a certain amount of legal autonomy to groups whose religious faith differed from that of the regime.[14]

At the other end of the Muslim world, the administration of British India permitted significant legal and governmental autonomy to the tribal groups of the far northwest of the sub-continent. The Pakistani government—often because it is *unable* to exert power in this area—has prolonged this arrangement, in today's Federally Administered Tribal Areas.[xv]

Perhaps one of the most interesting examples of this tolerance of indigenous legal systems is the patchwork of laws and rights that govern the many Native American reservations in the U.S. During the U.S. expansion across the country to the West, the U.S. government signed many treaties with the Native American groups that it defeated. These treaties not only gave the Native American groups rights over certain lands, but also gave them sovereign rights to self-government. As a consequence, today the U.S. could be considered a nation state with mini-Native American states scattered within it.

This arrangement has resulted in an unusual dual legal system on most Native American reservations. Although all Native American groups are subject to the national laws governing the U.S., while on their reservations Native Americans are *also* subject to their own indigenous laws. Each Native American tribe can regulate, judge, and punish their own members according to their own cultural and religious traditions.

Complicating the picture is the issue of the uneven and inconsistent agreements that were made by the U.S. government with each individual tribe. Since treaties were made with different tribes at different times in history by different U.S. representatives, few treaties are alike. The result is a confusing array of legal relationships between the U.S. and the hundreds of tribes scattered across the country.

Legal Systems within Legal Systems – The Native American Case; continued.

In recent years, a number of Native American tribes have used their political and legal autonomy to their own advantage. Thus we find that while many U.S. states forbid the sale of fireworks and gambling, many Native American tribes permit these activities, attracting much business particularly around the Fourth of July.

The Ottoman case demonstrates the unique legal and regulatory arrangements when the dynastic regime does not have a tradition of enforcing its faith-based legal writ over far-flung, religiously diverse populations. The Pakistani situation case emerges when the over-arching state lacks the legitimacy or coercive power to enforce control far from the capital. By contrast, the Native American case illustrates the complex and challenging problems that even a strong, organized state faces in dealing with a legacy of conquered indigenous peoples. Marines can expect that such problems only multiply in weak states whose power and authority over indigenous populations is marginal at best.[16]

The disconnect between state law and local law and practice is by no means total, however. In most societies, the state's legal system is significant, in cultural terms, because it reflects larger principles based on that society's norms, mores, and taboos. For Marines, the **implications are three-fold. First**, a Marine must work to grasp the norm, or more, behind the explicit law of the land. **Second**, a Marine will find that it is not enough to ensure that actions align with law; in a cultural sense, they must also adhere to local norms, mores, and taboos in order to obtain that legitimacy often necessary in the kinds of operations Marines undertake among foreign peoples.

Third, though U.S. or coalition forces seeking to establish stability in the AO may establish laws, decrees, or procedures, if they do not adhere to the local norms and mores, people are unlikely to follow these external laws and decrees. Though resultant indigenous behavior might be "illegal," that does not imply that the local people are lawless—rather, the externally-imposed legal order has inadequate legitimacy according to local norms and mores.

Culture Operator's Questions: Norms, Mores, and Taboos

In preparing for work in a foreign AO, Marines should try to determine answers to the following questions:

- What food and behavioral taboos exist in the region?

- What norms should Marines, even though they are foreign to the AO, observe?

- What underlying allegiances or codes of honor could influence the success of an operation?

- What activities in the area are considered serious violations of social mores and could carry serious punishments, including death?

- What beliefs or assumptions exist locally about American practices as regards local norms, mores, and taboos?

- What might the local people think (or have been propagandized to think) Marines are likely to disregard in terms of local norms, mores, and taboos?

Religious Beliefs

It is not uncommon for most people to assume that cultural beliefs are the result and reflection of religious beliefs—that religion determines the culture. Yet frequently, the reverse is true: that religious beliefs are adjusted to fit into local cultural beliefs—that cultural beliefs and practices influence the manifestation of religion.

In fact, particularly in areas without the mass communications, printing presses, etc., which can overwhelmingly influence belief systems, "syncretism" emerges quite often. **Syncretism** is the amalgamation of local cultural traditions, indigenous religious beliefs, and formal religious systems to create a synthesis of beliefs and traditions, which, when compared to the "proper" religion of scriptures, clergy, and capital-city seminaries, looks "strange" and "incompatible" with the "real" religion.

Example: Syncretism as Social Reality

A classic example of syncretism is the practice of Catholicism by the Indians of Peru. In the high Andes, Catholicism (which arrived with the Spanish) is mixed with ancient Native American traditions based on the Incan calendar and seasons. In a mix of Catholic religion with pre-existing local cultural practices, prominent figures in the ancient Indian religious stories and traditions suddenly become revered Catholic saints. These "Catholic" saints are worshipped by local Highland Quichuas, with their own saint days and festivities. Naturally, the Vatican does not recognize these saints, but has had little to no success in demoting the local religious figures in the local Peruvian communities.[17]

In discussing culture and religion, most anthropologists distinguish between the "Great Tradition" (the formal, written canonic version of a religion) and the "Little Traditions" (the local, informal, daily practices of a religion, which vary from region to region and even community to community).[18] In working in an AO, Marines need to be able to distinguish between what people in a community *do*—their daily observance and practice of their religion—and what the formal priestly religious authorities say that a "good Muslim or Christian or Jew or Buddhist" *should do*. As most Marines with experience in a foreign AO will attest, there is often a huge gulf between the two.

In fact, people from an AO reporting to Marines about religious practices and beliefs there will often tell Marines what people *should* do, as opposed to what they *do* do; conversely, often what people *do* do is passed off as the proper *ought-to-be-done*. Marines need to understand that in any given AO, both dynamics are underway.

As important, Marines need to recognize that local practice (rather than formal religious edict) ultimately has much more power to influence local peoples' behaviors and beliefs. For this reason, before deployment to a new area, it is not enough to study the "religion" as understood in texts, rules, or sacred histories of the orthodox community. Rather, it is **through studying religion as locally lived human performances and a locally manifested cluster of beliefs and attitudes that a Marine can come to understand religion** as s/he will see it and as people feel it in an AO.

Recognizing the gap between the formal and informal practice of religion is also immensely instructive in illustrating areas of potential tension and conflict. Below we briefly describe the important features of both formal and informal interpretations of religion:

Formal Religion. Formal interpretations of religion (regardless of whether the religion is Christianity, Confucianism, Islam, Buddhism, Judaism, Hinduism, or any other major world religion) are typically characterized by the following features:

• Texts, religious debates, and commentaries are written and passed down over time.

• Study occurs in religious schools that frequently (although not always) teach an officially composed and accepted doctrine.

• "Ownership" of the religion rests with an elite educated class of religious specialists (priests, rabbis, ulama, monks etc.) who are authorized to provide "official" interpretations, laws, and religious proclamations for the "flock."

• Differences between religious leaders and schools may lead to formally recognized schisms within the Great Tradition (Catholicism versus Protestantism; Shi'i versus Sunni Islam).

• The general populace has little role in the leadership, interpretation, or direction of the formal religious system; it is guided, but does not guide.

Informal Religion. Mingled with but in contrast to formal religion, informal traditions are predominantly folk interpretations of the Great Tradition. By "folk" we mean those developed through the prism of popular practices, beliefs, aspirations, and needs. "Folk" religion emerges outside formal, clerical, or "orthodox" structures, and its authenticity thus rests upon its resonance with the people, as opposed to textual justifications. Part of this authenticity, of course, reflects "folk religion," or the syncretistic mixture of indigenous beliefs, traditions, and practices found in Little Traditions. Marines will encounter these Little Traditions. These practices and beliefs will differ sometimes from town to town, and will factor into indigenous conduct towards Marines.

Little Traditions are characterized by:

• A predominantly mouth-to-ear method of inter-generational transmission of information about the religion.

• Hyper-localized variety in religious practice and interpretation, as the oral information becomes connected to other local traditions and beliefs.

• Religious prestige mingles with existing, often hereditary social structures, as folk religious leaders inherit position and knowledge from relatives or mystical teachers.

• Religious leaders become vested with almost superhuman powers, since their authority is based on both "secret" oral knowledge not obtainable through formalized learning methods, as well as on lineage relationships.

• Religious interpretations and daily practices are frequently based upon local traditions, beliefs, superstitions, and taboos, all reinforcing existing cultural norms.

• Power and authority are based on popular support and charisma.

• Local popular religious movements (such as the early Sufi movements in Islam, or Hasidism in Judaism) are perceived as challenges by the formal religious authorities.

Example: Folk Religion as Religious Transgression

Expressions of religion in many areas are often written off by people who say, "That's not the religion, it's only the culture." It is important that Marines see what goes on in front of them, however, as an interpretation of the religion in the local culture in a way that creates a meaningful spiritual and social existence. In monotheistic religions, amulets exemplify this dynamic.

From Kenya to Senegal, many African soldiers wear amulets under their utilities for protection against natural and man-made violent forces. This inanimate object is believed to have the power to affect humans and nature. For soldiers who are Muslim—or Christian—this is prohibited in the formal, orthodox faith, as it implies that something other than the omnipotent God influences natural events and men's actions. Yet, soldiers who are Muslim see no disconnect between the use of amulets and the Islamic prohibition against worshipping any object or spirit other than God; instead, they see it as part of their Islam.

Lest this be thought the case only for primitive, semi-monotheistic Africans, throughout the Eastern Mediterranean and Middle East, Christians, Jews, and Muslims wear amulets such as nazar charms (eyes), hamsas (stylized hands), or boxes with mystical parchment in them to ward off the evil eye.

Though none of this is sanctioned by the canonical texts of religion, even "orthodox" members of the faith use them, with related expressions prevalent in the spoken, even secular languages. Jews say "bli 'ayin ha-ra'" (without the evil eye); Muslims, Christians, and Jews say "ma sha'llah" (as G-d wills it [so the evil spirit will not taint it]), etc.

In fact, much of performed religion that Marines will see in an AO is part of the Little Tradition, proscribed by "orthodox religion." Beating oneself on 'Ashura is frowned upon by Shi'ite religious authorities trained in the seminaries—just as clergy in the Philippines forbid—but look the other way—when Catholics nail themselves to crosses on Good Friday. Ecstatic, continuous bowing during prayer—visible in any traditional Jewish synagogue service—is considered improper decorum in strict commentaries, while the whirling of Sufi dervishes is thought libertine and excessively sensual by the Islam of the ulama. But it continues, because it speaks to peoples' needs and often is layered upon pre-existing norms and practices.

Remember: there is what is *supposed* to be—believed, performed—and then there is what really *is*. Often, what *is* has more meaning for people who actually live the religion on a daily basis than what is *supposed* to be. Similarly, belief and creed systems are not sealed off from each other. Grasping these realities will allow ground level operators as well as information operations personnel to effectively incorporate these beliefs and practices—all narratives—into their operational planning.

Culture Operator's Questions: Religious Beliefs

Questions that Marines can use to guide their evaluation of the role of religion should include:

- Who is the actual leader of the local religious community?

- How do religious leaders relate to the educated elite vs. popular groups, etc.?

- What is the basis of authority for a "religious" leader in the AO: book learning, lineage, charisma, etc.?

- What are the actual (versus theoretical/textual) religious practices in the specific AO where the Marine operates?

- How do local practices of a religion the Marine has encountered elsewhere differ from what that Marine thinks the religion is "supposed" to look like?

- What power and role, if any, does the formal religious system play in local peoples' daily lives?

- What conflicts or disagreements exist between the formal religious system and the local religious practices of the AO?

- How prominent is "religion" as an explanatory factor for people in current events, and in reference to history, or historical trajectories?

- What is "the way the world is supposed to be" according to locally-held religious beliefs, and how does the Marine presence impact that?

Notes

[1] "Group beliefs" should be distinguished from "individual beliefs." The former are open to influence by behaviors and structured relationships; the latter are much less so.

[2] Maurice Halbwachs, *On Collective Memory* intro. and ed. By Lewis A. Coser (Chicago: University of Chicago Press, 1992); also see discussion in the introductory chapters of Yael Zerubavel, *Recovered Roots: Collective Memory and the Making of Israeli National Tradition* (Chicago: University Of Chicago Press, 1997).

[3] For a discussion of this in the context of Egypt, see Barak Salmoni, "Historical Consciousness for Modern Citizenship: Egyptian Schooling and the Lessons of History under the Constitutional Monarchy," in Goldschmidt, Johnson, Salmoni, eds., *Re-Envisioning Egypt, 1919-1952* (Cairo: American University in Cairo Press, 2005), 164.

[4] Hamid Dabashi, *Theology of Discontent: The Ideological Foundation of the Islamic Revolution in Iran* (New York: Transaction Publishers, 2005); Vali Nasr, *The Shia Revival: How Conflicts within Islam Will Shape the Future* (New York: W. W. Norton , 2007); Yitzhak Nakash, *Reaching for Power: The Shi'a in the Modern Arab World* (Princeton, NJ: Princeton University Press, 2007); Ali Allawi, *The Occupation of Iraq: Winning the War, Losing the Peace* (New Haven, CT: Yale University Press, 2007).

[5] See Jim Proser and Jerry Cutter, *"I'm Staying with My Boys..." The Heroic Life of Sgt. John Basilone, USMC* (Lightbearer Communications Company, 2004).

[6] Personal observation by author, RCT 5 HQ, Camp Falluja, Iraq, October 2006.

[7] Zdenek Salzmann, *Language, Culture, And Society: An Introduction to Linguistic Anthropology* (Boulder, CO: Westview Press, 2006).

[8] Douglas McCollam, "The Umbrella's Shadow," *Foreign Policy* (online), October 2006; also see Jack M. Holl's comments about the umbrella's symbology into the 1950s in American political culture: "The New Washington Monument: History in the Federal Government," *The Public Historian* 7:5 (Autumn 1985), 9-20; also see the excellent recent book by Jeffrey Record, *The Specter of Munich: Reconsidering the Lessons of Appeasing Hitler* (Dulles, VA: Potomac Books, 2007).

[9] Technical terms for these ideas include "haptics," "tesics," "proxemics," among others. For more on this topic in general, see Muriel Saville-Troike, *The Ethnography of Communication: An Introduction* (Boston, MA: Blackwell Publishing Limited, 2002); Joy Hendry, *An Anthropology of Indirect Communication* (London: Routledge, 2001).

[10] BN CO from 1st MarDiv, interviewed by author, OIF II, TCMEF, April 2005.

[11] For a Lebanese example of this dynamic, see, Augustus Richard Norton, "Ashura in Nabatiya," *Middle East Insight* 15:3 (May-June 2000), 21-28.

[12] Center for Advanced Operational Culture Learning, *Afghanistan: Operational Culture for Deploying Personnel* (Quantico, VA: Marine Corps Training and Education Command, 2006), 51-61.

[13] See Hussain Haqqani, "Head Scarves and History: For Clumsy Secularism, Deadly Rewards," *International Herald Tribune* 22 Dec 2003; John P. Entelis, *Islam, Democracy, and the State in North Africa* (Bloomington, IN: Indiana University Press, 1997); Fred Halliday, "Tunisia's Uncertain Future," *Middle East Report* No. 16 (Mar-Apr 1990), 25-27; "The Art of Plain Talk," *Time Magazine* 29 Sept 1967.

[14] See Halil Inalcik, "Ottoman Methods of Conquest," *Studia Islamica* 2 (1954); idem., *The Ottoman Empire in the Classical Age, 1300-1600* (London: Weidenfield and Nicolson, 1973); Colin Imber, *The Ottoman Empire, 1300-1650: The Structure of Power* (London: Palgrave-Macmillan, 2004), 216-51; Benjamin Braude and Bernard Lewis, *Christians and Jews in the Ottoman Empire: The Functioning of a Plural Society* (Princeton, NJ: Holmes and Meier, 1982); Ira Lapidus, *A History of Islamic Societies*, 2nd ed. (London: Cambridge University Press, 2002). For the dilemmas that millet-like arrangements created for modern Middle Eastern states, see Elie Kedourie, *Politics in the Middle East* (London: Oxford University Press, 1993).

[15] For journalistic coverage of the Pakistan and its Federally Administered Tribal Areas (FATA), see Mary Weaver, *Pakistan: In the Shadow of Jihad and Afghanistan* (New York: Farrar, Straus, and Giroux, 2002); as well as Zahid Hussain, *Frontline Pakistan* (New York: Columbia University Press, 2007); also see Benedicte Grima, *Secrets from the Field: An Ethnographer's Notes from Northwestern Pakistan* (Bloomington, IN: Authorhouse, 2004).

[16] See N. Bruce Duthu and Colin Calloway, eds., *American Indians and the Law* (New York: Viking Press, 2008); Robert N. Clinton and Carole E. Goldberg, *American Indian Law: Native Nations and the Federal System*, Fifth Edition (Matthew Bender: 2007).

[17] Michael D. Sills, "Christianimism: Catholic Syncretism in the New World," *Southern Baptist Theological Seminary*, http://www.sbts.edu/pdf/ChristiAnimism.pdf; Dora G. Lodwick, "Reviewed Work(s): *Peru: When the World Turned Dark*," *Teaching Sociology* 19, number 1 (January 1991): 117-119.

[18] For descriptions and explanations of Great and Little Traditions in the context of formal versus informal religious systems see Clifford Geertz, *Islam Observed: Religious Development in Morocco and Indonesia* (Chicago, Ill.: University of Chicago Press, 1968); Dale F. Eickelman, *The Middle East and Central Asia: An Anthropological Approach* (Upper Saddle River, NJ: Prentice-Hall, Inc, [1981] 1998).

Part III

Toward Applying Operational Culture

In Part I of this book, we described and defined the basic features of culture in terms of their relevance for the Marine. In Part II we moved from the more abstract concept of "culture" to five specific dimensions of culture that affect Marine operations, providing examples from different regions of the world where Marines have operated and may deploy. Ultimately, however, theories and principles of culture are only relevant for the Marine if they can be applied to actual operations.

It is not the purpose of *Operational Culture* to provide a doctrinal manual with specific steps that Marines are expected to follow in applying culture to military operations. Indeed, to the contrary, throughout this book we have emphasized that Marines must use principles of culture in a flexible and creative manner in order to achieve operational success. There is no one formula or checklist that can serve Marine purposes across the spectrum of military activities around the world. "One size fits all" does not suit the application of culture to the spectrum of Marine involvement in foreign AOs. Just as operational and strategic planning requires an intuitive understanding of the process, incorporating culture into operations is as much an intellectually-informed "art" as it is a "science."

In Part III, therefore, we turn our discussion to the "art" of applying principles of culture to Marine operations and activities. Chapter Eight applies the Five Operational Culture Dimensions to three case studies: Nigeria, Darfur and the Philippines. In each case, we examine the way that all five dimensions interact to create specific conditions in the region, and discuss the impact that these sociocultural conditions could have on potential Marine operations in the region. Chapter Nine then turns to the issue of teaching cultural principles to Marines, both in Professional Military Education (PME) and during pre-deployment training (PTP). In order to provide an effective framework for military education, we use Bloom's Taxonomy of learning to evaluate the appropriate kinds and levels of culture learning necessary at different ranks, billets, and functions.

Chapter 8

From Models to Dimensions
to Observable Realities

In Part II of *Operational Culture for the Warfighter*, we examined the three core models of operational culture, and then we moved on to the five dimensions through which a culture operator can understand, plan for, and work in an area of operations. We examined the ways these dimensions have practical and concrete relevance to Marine operations by discussing the components of the various dimensions and the relationships that connect them.

"Relationship" is the focal word here: in culture groups, none of these models, dimensions, or components exists in isolation from one other. Rather, they are all integrated. In every case, a culture group's realities in one dimension set the stage for conditions in the next dimension; while for some of these dimensions, the cause and effect relationship goes in two directions. The same is true for the components of the Five Operational Culture Dimensions.

It is essential to grasp the integrated nature of cultural dimensions in order to fully understand and take advantage of cause and effect within an AO. This is not unlike the combined operations of a Marine Air-Ground Task Force (MAGTF). The presence of the Wing, Combat Services Support, Logistics, Ground Combat Element, and Intelligence, Surveillance, and Reconnaissance (ISR) assets is not enough in and of itself. Each can interrupt the lines of operation of the other if they are not fully integrated so as to be mutually supportive in pursuit of a common operational or theater-level goal. It is the role of command, control, and communications—C3, and of course, good commanders—to provide such integration and ensure proper battle management.

In the same vein, in the absence of an integrated understanding of culture, Marine actions, be they kinetic, civil affairs, information operations, or intelligence collection, will be contradictory and undermine each other. Most importantly, these un-integrated actions can often cause the opposite effect from the one intended on the local civilian population, opponents, and indigenous partners. This means that **an un-integrated approach to integrating culture in operations will undermine the mission itself: It will turn culture into a barrier confronting battle management.**

In Part I of this book, we introduced the Five Operational Culture Dimensions as emerging from three analytical models based on twentieth-century anthropology. The diagram below refreshes our memory of this. Just as important, the connecting lines demonstrate the interrelationships, or backwards and forwards lines of influence, among the Operational Culture Dimensions themselves. More broadly, the diagram shows that the Ecological, Social Structure, and Symbolic analytical models are tightly interlaced. This interlaced, recursive nature of Operational Culture should inspire a Marine's analytical methodology when examining cultures in the operational planning and execution phases.

Figure 8.1
Operational Culture: From Models to Dimensions

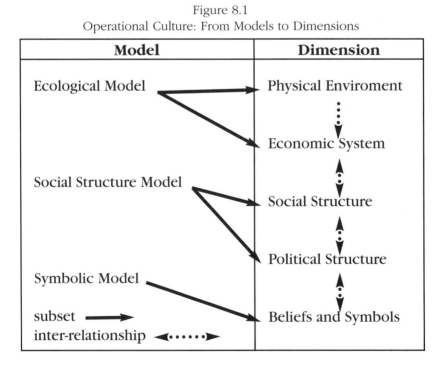

Part II described how each Operational Culture Dimension contains several components, which also emerge in the battlespace in dynamic interaction with each other. The diagram below depicts this relationship, both among Operational Culture Dimensions and their Components.

Figure 8.2
Operational Culture: From Dimensions to Components

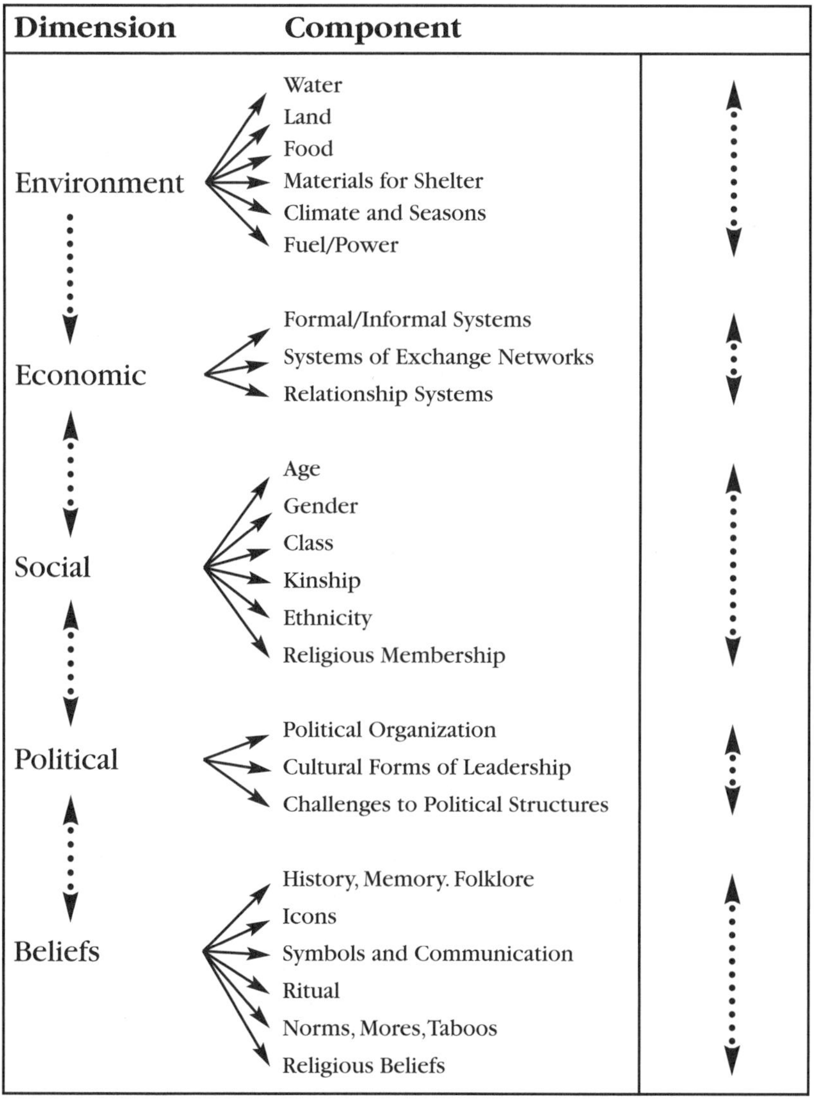

Dimension	Component	
Environment	Water Land Food Materials for Shelter Climate and Seasons Fuel/Power	
Economic	Formal/Informal Systems Systems of Exchange Networks Relationship Systems	
Social	Age Gender Class Kinship Ethnicity Religious Membership	
Political	Political Organization Cultural Forms of Leadership Challenges to Political Structures	
Beliefs	History, Memory. Folklore Icons Symbols and Communication Ritual Norms, Mores, Taboos Religious Beliefs	

These diagrams do not merely illustrate theoretical linkages; dynamics in historical and contemporary cultures bear them out. In this chapter therefore, we examine three case studies that show how in contemporary operating environments, cultural dimensions and their components are in fact tightly integrated, producing either the conditions for conflict, or the conflicts themselves to which Marines will deploy.

Case Study I – Nigeria: Oil and Burgeoning Insurgency

In the 1950s, Nigeria was like many other Central/West African countries emerging from colonialism, in this case British. It was poor by Western standards, and possessed an economic system based on agriculture. In the Niger Delta in particular, an environment of intense wetlands, creeks, rivers, and mangroves facilitated cultivation of palm oil and cacao beans, while fishing was prevalent as the Delta opened onto the Atlantic Ocean. This environment permitted an economic system supporting a relatively self-sufficient subsistence based on cultivation and trade of local foods, with barter playing a role equal to that of currency. Currency in circulation locally tended to be used more for purchase of survival commodities, as opposed to capital accumulation, investment, or development of capacity for migration or absentee forms of ownership.

The environment therefore provided economic livelihoods to small communities whose social structures were in turn organized through kinship systems. These kinship systems—or tribes—were also associated with hyper-local ethnic groups, such as the Ijaw, Igbo, Igoni, and Osoko—often labeled by previous colonial rulers.

These groups, sometimes tribal, sometimes ethnic, identified closely with their local territories, though they were citizens of a political state whose boundaries were drawn by exiting colonial rulers—and which included other, majority ethnic groups such as the Yoruba, Hausa, and Fulani, who were from the northern, more Muslim part of the new state. In the Delta, therefore, social structures producing political structures of hereditary tribal leadership resulted in local identities, though local kinship network-based rulers possessed much less coercive or economic power than did the central government, dominated from the 1960s by the military.

This "cultural ecosystem"—linking environment to economy, social structure, and political structure—began to evolve from the early 1960s, when oil began to emerge from the Niger Delta in commercial quantities. In the ensuing fifty years, though Nigeria has become an oil giant of strategic significance to the U.S., Europe, China, and Japan, the economic changes have driven political, environmental, and social structural deformities throughout Nigeria.

Spanning across all Five Operational Culture Dimensions discussed in Part II, these deformities are most evident in the Delta itself, resulting in communal conflict and proto-insurgent conditions, in turn threatening U.S. and Western strategic interests through potential military conflicts and shocks to industrial economies reliant on African oil.

Oil is often considered an economic commodity, which it is. However, it is in the first instance a component of the natural environment—and its uses will thus have significant impacts on the environment as well as the economy of a culture. As we have suggested, not only does the economy of a culture influence its social structures, but political developments can ripple through the social, economic, and even environmental dimensions of operational culture. This has been the case in Nigeria over the past fifty years.

The discovery of oil in commercial quantities in Nigeria came at roughly the same time that post-colonial Nigeria experienced a series of military coups by officers who had been trained by the British during their imperial control of the region. In the main, the coups were motivated by the desire for power and control of resources, but also reflected competition among the three Islamized majority ethnic groups from the north of the country, the Yoruba, Igbo, and Hausa-Fulani. For none of these were the minority ethno-religious groups of the Niger Delta a priority.

These military rulers, from 1966 through 1999, were to an extent also influenced by currents of third-world nationalism. Here, natural resources—such as oil—were the property of no individual or sub-region, but rather were the patrimony of the government-as-nation. In 1971, the military-led Nigerian government nationalized the oil industry, so that all revenues from oil export come to the government, which would then dole out revenues to regions. In effect then, an aspect of the environment with economic implications became part of both politics as well as ideological belief systems, such that political agendas influenced the economy.

Like other oil-rich states, Nigeria prioritized the oil industry at the expense of more traditional agriculture and hunting and gathering—practices that put less of a stress on the environment and were also able to sustain the economic and social structures of small chief-led communities in the Niger Delta. The oil economy has physically deformed the environment of the Delta. Mangroves, in whose swamps fish collected, have been stripped away so that oil pipelines could criss-cross the Delta. The pipelines, as well as new roads ploughed through marshlands, have reduced areas for fish to spawn.

Like their food for survival, fishing communities themselves have been physically uprooted and moved to other parts of the Delta, to make room for fuel storage tanks and natural gas plants. In these same areas, forests have been uprooted, and no longer provide natural wind and rain blocks for local communities. Families thus find their thatched roofs rained through and ruined much more frequently, with replacement becoming increasingly costly. Also, the burning oil flares have released greenhouse gasses, heating up the climate and causing acid rain that corrodes roofs, hurts crops, and leads to respiratory illnesses.

Finally, the increased large vessel traffic and terminal construction on the Delta coast have changed wave patterns, resulting in soil erosion. Remaining fish stocks are thus found farther out to sea than in the past, and few Nigerians on the coast can afford boat engines powerful enough to permit them to reach their traditional sources of food and economy. The agricultural and fishing economy is now swept away, along with the environment that supports it—some towns have seen their populations drop by 90% in the past 30 years, while Nigeria is now forced to import food. If a Deltan wants fish, he must purchase frozen fish shipped in from abroad—if he can afford it.

The new economy is the oil economy, and it has influenced the local social and political structures. The government in Abuja, even after the return of civilian rule in 1999, has prioritized areas outside the Delta for expenditure based on oil revenues. At the same time, it has helped the dominant tribal and ethnic groups—Yoruba, Hausa-Fulani, and Igbo—to find jobs in the Delta's oil economy, thus economically excluding the local peoples as a corollary of their political exclusion.

Further, the new economy impacted the local village- and tribal chief-based political structures. The Abuja government, as well as foreign oil companies, have offered irresistible amounts of money to local chiefs in order to obtain drilling rights—rights which are traditionally not those of the chief to grant unilaterally, given the social structural dynamics of tribalism as seen on Chapter Five. Other, economically depressed chiefs without agricultural or fishing resources, are now able to remain well-off by supporting the oil economy through road-building crews or sand-dredging.

In both cases, social status and political power of chiefs are no longer based on a now-weakened social compact with a kin group; rather, local leaders are reliant on the government and foreign economic agents who produce and extract oil. Chiefs are thus weaker than before as nodes of socio-political control and local mobilization.

What remains? Unemployed people in the fifteen-to-fifty age bracket. Disenfranchised politically, economically, and ethnically by the state, they constitute a sector of society which because of its mere positionality in Nigeria, is bound to challenge the existing political structure. They confront much weakened local social and political structures, therefore unable to restrain them.

In this environment, new belief systems have become popular, including local versions of Protestant Christianity—adding another complicating element to the north-south ethno-religious divide. New belief systems also emerge through social structures, such as church communities and preachers, whose moral suasion might rival that of tribal leaders on the decline.

These new social groups also have their historical memories and icons: In 1990, the writer Ken Saro-Wiwa founded the Movement for the Survival of Ogoni People. The organization demanded control of the oil on local lands and an end to environmental damage. In 1993, a quarter of a million locals rallied in support. The military government retaliated by charging Saro-Wiwa and others with murder on ethno-tribal bases. Saro-Wiwa and eight others were hanged by the state in 1995.

Such events, and the beliefs and symbols they generate, have set the stage for the most threatening new development in the social and political structure: a local insurgency. The Movement for the Emancipation of the Niger Delta and other lesser-known groups have emerged, led by people much lower in the traditional social scale, with less education, and at times from traditionally less prominent branches of tribes. These groups have in the past few years attacked local oil infrastructure, local police and military forces, and representatives of American and European oil companies.

In the latter case, the Movement and shadier groups can tap into the memory of Nigeria's colonial history to tar the companies— and the Abujan government—as complicit in re-instituting imperialist control of the nation's local resources. Ironically, rebel groups can use against the government an ideological axis—control of the nation's natural resources against foreigners—that the post-independence political regime itself had made popular.

The Nigerian army has responded by destroying villages, expelling Deltans from Military Zones, and engaging in mass round-ups of young men. In the process, international oil companies have been forced to reduce work—also harmful to the local economy now dependent on French, Italian, and American firms—while Western countries themselves have begun to consider the implications of Nigerian Delta insurgency for strategic stability and terrorism.[1]

Considering this case in light of the Five Operational Culture Dimensions studied in Part II of this book, we see that an insurgency of concern to Western governments today results from the highly integrated nature of environment, economy, social structures, politics, and belief systems. The economic changes, in conjunction with political changes at the national level in Nigeria, altered the local environment, economy, social structure, and political dynamics in the Niger Delta. Likewise, these changes were reflected in new ideologies, methods of mobilization, and ultimately new and dangerous political-military processes. This all threatens stability in a key African state, potentially endangers energy supplies for the West, and might at some point require US intervention.

Without considering these local cultural dynamics in an integrated fashion, a Marine commander will be unprepared to deal with the challenges and opportunities he will face. This is because he and his Marines will likely encounter local leaders who are of questionable residual legitimacy, and cannot guarantee the loyalty of "their" people. If intervening to support foreign oil companies, Marines will automatically be associated with the imperial legacy and foreign companies in collusion with an alien ethnic, religious, and political regime from Abuja.

Marines will also find military-aged males, whom it is inappropriate—in an operational culture sense—to label as "good," "bad," or "evil;" rather, their mere positionality in the social, economic, and political structures has pushed them into a destabilizing role. These Niger Deltans will have suffered injustices at the hands of other uniformed people—the Nigerian military—and will thus feel quite justified in engaging in destabilizing activities.

Conversely, though some of the leaders of these groups might be able to enforce quiet among their followers and ethno-religious kinsmen in areas of operation where their kinsmen form majorities, the Nigerian state or foreign oil companies would be likely to oppose the granting of legitimacy to these "terrorists."

Further, likely local partners in stabilization or security operations will be police and army leaders who are ethnically and religiously alienated from the communities being secured and stabilized—they will also be strikingly corrupt, as an oil economy in an authoritarian state turns representatives of the state into extractive agents in society. It will also be a military that is somewhat ethnically and geographically divided, and does not trust itself, given its memory of spawning several military coups and counter-coups.

Other possible partners—the UK, France, or Italy—might be associated in local minds with the foreign oil companies that have been complicit with the regime in wrecking environment and economy.

A task force commander and his subordinate commanders will also encounter severe humanitarian and civil affairs needs, from medical problems to malnutrition, from low levels of education to unemployment as the rule. Providing funds to village leaders to see to these needs will likely result in little amelioration of conditions. These leaders have less sway over communities than an American is likely to assume a tribal "leader" would have. Further, these local leaders have become accustomed to social and political structures where the state or foreign oil companies dole funds out to them, but do not hold them accountable for the conditions of their villages.

Finally, returning to the environment, because of soil erosion and weakening of terrain and ecosystems, speedy mounted movement might be difficult, except by negotiating use of roads for fees with locally compromised leaders—or insurgent bands, whose notions of honor, law-and-order, etc., are distinctly different from those of traditional local leaders. Further, extended amphibious operations will aggravate ongoing environmental destruction, as will mounted patrols in tracked vehicles, seven-ton trucks, and other heavy equipment currently in use in urban contexts. In these vehicles, Marines will be seen as part of the problem.

A Marine commander without this understanding cannot function effectively as the "culture operator" discussed in Chapter Two. However, equipped with the analytical method laid out in the past several chapters, he can plan and equip his force appropriately; mitigate false impressions about his force; locate proper local points of contact and influence; conduct operations in a way lessening impact on the local people; and begin to provide for some of their needs on the environmental, economic, and political levels. The commander, in this case, can also in a more informed fashion balance concerns about culturally appropriate conduct with force protection and use of kinetic effects, all in the service of achieving the task force's mission end state.

Case Study II – Darfur: Environment, Economy, Ethnicity, and War

Darfur is a region of western Sudan. It has been the site of a local insurgency, and government-backed or government-run counterinsurgency, since 2003. It is often cast as an Arab ethnic cleansing campaign against Africans. The realities and roots, however, go back to the 1980s. Here too, the Five Operational Culture Dimensions have been at play.

Up until the 1980s, the physical environment of Darfur had been divided between fertile lands and rocky outcroppings, hillsides, and ridges. Though not at Mediterranean or European levels, rainfall was seasonally constant, with water absorbed by plants and alluvial soil. Therefore, while not supporting modern communications or transportation—and thus urban industrial economies—this physical environment did generate two dominant modes of economic life, co-existing symbiotically. One of these was agriculture, the other pastoral nomadism.

On the fertile plots, farmers cultivated crops that supported a subsistence economy, providing for regional peoples' needs when rainfall was adequate. On the hillsides and crags separating the cultivable plots, nomads grazed their animals. The two economic lifestyles interacted, as farmers and nomads shared wells, with the former purchasing goods and animals from the latter, and the nomads gathering in harvested crops that the agriculturalists left behind for them, as part of a conscious tradition of symbiosis.

Darfur's social structure is revealed in the meaning of the word "Darfur"—"land [or abode] of the Fur," with *Fur* being the largest agriculturalist tribe in the region.   Both nomads and farmers possessed social relationships based on tribes, with these tribes being associated with particular geographic—and thus agricultural or migratory—sub-regions, or *dars*, in the area.   Social structure was thus kinship-based, with identities revolving around three related axes—economic mode of life (farmer or nomad), tribe, and lands ancestrally associated with that tribe (*dar*).

Ethnic labels also exist, but they related to economic and territorial realities and not ehno-religious divides or a sense of difference. Rather, to be "black" or "African" meant to be a settled farmer; to be "Arab" meant to be a nomadic pastoralist.   In the same way, "African" areas were no more than the *dars* of the settled farmers; the "Arab" areas were the *dars* of the nomadic pastoralists. "Africans" and "Arabs" are not racially distinct in Darfur, though there are language differences.   Significantly, then, environmental parameters produced modes of economic life, which were reproduced through social structures that sustained a system of identities.

It follows then that changes to the environment would produce alterations to the way other Operational Culture Dimensions manifested themselves in Darfur.   Along with political dynamics in Khartoum, it was these changes—and not "age-old cultural animosities" between "Arabs" and "Africans," or desires to "ethnic cleanse" Africa for the sake of "Muslim Arabs"—that produced and prolong the Darfur crisis.

In the late 1970s, seasonal rains began to fail in Darfur.   Farming therefore became harder, sand blew into once-fertile lands, and existing alluvial soil was washed away by the rare rains.   "African" farmers began to fence off lands once open to the herds of "Arab" nomads, and did not supply them with agricultural produce.

As the environment became incapable of supporting two economic systems, nomads had the choices of flight to equally drought-ridden areas; selling off herds and taking up farming on marginal plots—against the grain of their social identities as free nomads living in pastoral *dars*; or raiding and encroachment into the *dars* of sedentarized agricultural tribes. In increasing instances, "Arab" nomads chose the latter: The environmental change spurred economic conditions forcing lifestyle choices that interacted with cultural identities to create political problems and violence.

In the 1980s, groups of landless nomads began to raid and occupy "African lands." The political and military leadership in Khartoum, without any roots in Darfur or relationships to agriculturalists there, encouraged the nomads to think of themselves as ethnically Arab, fighting against the Africans who unjustly kept them from lands and sustenance. In the late 1980s, ideologically encouraged by the state, the "Arab" nomads attacked the farmers, provoking farmer-nomad clashes in greater numbers.

On both sides, new social structures such as marauding bands emerged, as did political structures, with leadership at times drawn from sons of traditional tribal leaders or offshoots of traditionally significant families. Political manifestos encouraged by the state were paralleled by networks of nomads based on tribal and ethnic lines reaching to Khartoum, in order to appeal for and receive support from the political leadership there.

As more farmer villages and nomadic camps were burned in conflicts related to environment and economic challenges, both sides began to see themselves—and be seen by both the Khartoum elite and the outside world—in ethnic, racial, and sometimes religious terms. Fighting continued into the 1990s, so that by this time people on both sides began to "remember" "age-old" rivalries not only on tribal lines, but ethno-tribal lines as well. When nomadic tribes in Darfur found that the state would be responsive to "Arab" complaints, they advertised themselves as Arab, and encouraged their tribal members to do the same. In 1994, the government in Khartoum redistricted Darfur into three sub-regions, putting people from nomadic "Arab" kin networks in power. The new administrators used their power to territorially and economically marginalize the agricultural "black African" tribes, such as the Fur, Zagawa, and Masalit.

In 2003, as the north-south civil war drew to an end in Sudan, a rebellion broke out in Darfur. This was an insurgency led by the settled farmers in the "African" *dars* against a state they considered neglectful of the region, and supportive of the nomadic groups raiding the agricultural areas—to the point of turning a deaf ear to farmers' complaints; ideologically inciting the "Arab" nomads; and providing arms to them.

The Sudanese state then launched its counterinsurgency, in the main employing proxy fighters—the poorer, most dispossessed nomad tribesmen—organized into a militia known as the Janjaweed, based on a social and ethnic idea of "Arabness" against "black Africanness." A new, state-created political structure had thus come into being along with a related social identity, and for the first time in the long course of Darfuri history, geographic, social, and political relationships came to be perceived by protagonists as being drawn along ethnic lines. That this was a new phenomenon, artificially grafted onto previous economic relationships and social identities, can be seen from the fact that the "Arab" tribes who had suffered least in terms of land possession and economic livelihoods were also those least interested in being Janjaweed.

For their part, the "African" insurgents also developed new social relationships and networks within groups, at the same time as new political structures emerged in the form of political groups-cum-rebel militias. The foremost groupings have been called the Sudanese Liberation Army and the Justice and Equality Movement. Their leadership has somewhat followed geographic and tribal lines, as have their differences of opinion.

From the 1970s through 2006, therefore, environmental changes drove changes to the economic conditions, which restructured social relationships and caused political unrest. Economic and political difficulties, manipulated by the state, took on a racial hue, as people were encouraged, either directly by the state or in response to state actions, to assume ethnically-based identities.

These conditions then produced insurgency and counterinsurgency, where the underlying material interests that sparked the difficulties became layered over by ideology, ethnicity, and notions of tribal warfare. These circumstances also lead to complications at the international level. Given the Sudanese government's flirtations with Islamism and Bin Laden in the 1980s and 1990s, after 11 September 2001 it became very easy for Western leaders and activist groups to view the Darfur conflict through the prism of the War on Terror and Arab (Muslim) brutalization of "black Africans." There have thus been recurrent calls for intervention in Darfur to prevent ethnic cleansing and genocide against "African farmers" abused by "Arab Janjaweed," even though none of these categories really apply, and the "Africans" themselves began the insurgency.

At the same time, Darfuri agricultural tribes' refugee flight to their linguistic and ethnic kinsmen farther west has entangled neighboring Chad in the conflict. Chadian security forces have made intermittent forays into Darfur, and the government in N'Djamena has threatened economic boycott or harsher measures against Sudan in the name of security, though with an undertone of tribal and ethnic animosity. The United States has had a low-level though important security relationship with Chad for some years. It is also the east flank of the strategically significant Sahel region of Sub-Saharan Africa. This, combined with U.S. concerns regarding Sudan, Ethiopia, and then Somalia, might at some point make U.S. intervention in Sudan a possibility.[2]

If Marine forces were sent to intervene, they would likely have multiple missions: They would have to separate conflicting parties, interdict weapons smuggling routes, and in some cases use force to protect targets of attack. They would also need to dissuade Chad from intervening—or work with Chadians and other African militaries as partners. Commanders at all levels would also have to plan and execute sustained humanitarian operations, to include population relocation, medical care, physical reconstruction, building of shelters, and temporary job creation. They would need to do all this in collaboration with an unwilling Khartoum government, international aid agencies, non-governmental organizations, and residual African Union military forces.

A Marine commander, accustomed from Iraq and Afghanistan deployments to think that tribes have specific "AORs," and are also of specific ethnic and religious groups, would not be well served by these ideas. He would need to grasp that there are no static "tribal lay-downs" geographically, and that even his native interlocutors would be telling him only what they remember, or wished to be true. Rather, people, groups, and notions of "who's who" have been on the move over the last decade. At the same time, a Marine may well encounter groups of people who feel connected to parcels of land. These parcels could either be new areas to which they have been forced as displaced persons but now do not want to leave, or ancestral *dars* from which they have been ejected, but still consider "their" lands. These feelings of connectedness will be based on a combination of symbolic and economic value.

Further, Marines would have to understand that "African," "black," "Arab," and "Islamist" are often used as convenient labels not reflecting actual local causes of conflict. In fact, Marine commanders at all levels will need to detect initial material causes for the clash of needs and interests, providing inducements in the form of cash, humanitarian assistance, and reconstruction, to reduce the basic drivers of conflict—i.e., those things that will cause conflict no matter who is "African" or "Arab." Commanders will have to provide these inducements in such a way as to strengthen traditional kin authorities, along tribal and *dar* lines—thus sidestepping at times official Sudanese "Arab" authority chains—even if this means a fair amount of corruption and skimming by tribe leaders.

Corruption, theft, and unresponsiveness to measures which "objectively" would help Darfuris will at least partially reflect egalitarian economic notions about the need to spread wealth among many unseen segments of a group, so as to meet traditional welfare responsibilities. The need to maintain tribal leaders' prestige, or shore up their waning social esteem, will also drive corruption. Finally, since cultural groups exposed to prolonged arbitrary conduct by governmental powers often exhibit low levels of trust in both the future and anyone beyond close-in networks, there will be added corruption as well as skepticism or apathy towards Marine initiatives.

Still, Marines will have to determine when new labels have taken on meaning for protagonists in the conflict. In this case, a Marine leader will find that in addition to government incitement, sometimes enduring material interests—access to livelihood through land—cause violence, whereas at other times violence is prolonged because new ethnic labels take on a meaning of their own, and then create interest groups and conflict.

Furthermore, because ethnicity, tribe, and territory have become so entangled, a Marine will need to recognize that seemingly impartial U.S. actions which his forces take—separating combatants, arresting marauders, etc.—will be perceived by different protagonists as the U.S. taking sides and supporting one group over the other. Conversely, efforts to remain neutral and not intervene too aggressively on any one side will also be perceived by one or the other protagonists as bias.

Also, due to years of mutual bloodletting among people who mostly do not wear uniforms, distinctions among "civilian," "military," "combatant," and "neutral" will be irrelevant in local terms. Darfuris from one tribe, *dar*, or "ethnic" group will consider it legitimate to target unarmed members of other groups, as they are all potential enemies. Attempting to break this mental mold will be futile for Marines.

In the same vein, years of conflict will have produced a social stratum of male adolescents and young adults who have come to associate manhood with carrying a weapon. This will manifest itself in many legitimate "targets" being young by American standards, thus raising challenges for rules of engagement. Likewise, resistance to disarmament will be articulated as desire to guard local cultural understandings of manhood.

This will not be the only issue of gender relevant to Marines. Due to mutual raiding and state use of Janjaweed to attack settled areas, there will be many mothers whose husbands have been killed or cannot provide livelihoods. Likewise, it is possible that women who otherwise would be married have remained single due to fewer men to marry. They might engage in work that is considered culturally inappropriate to women in Darfur, requiring particular Marine awareness and precautions. Rape will have produced illegitimate births as well as social stigma for these women. Attempting to help them, find jobs for them, or regularize their legal status, might associate Marines with moral laxity.

Finally, a hypothetical Marine force will not be the only legal armed force on the ground. African Union troops, other Western forces, and perhaps the Khartoum government's own forces will be operating in the area, potentially cooperating in a coalition format, or perhaps coexisting through tacit understandings. Negotiating such formal and informal agreements will require the kind of interactions necessitating attentiveness to the symbolic meaning of words and actions.

Furthermore, no force from within the region or country will be "above the fray"—they will all have political, ethnic, perhaps even religious or class-based attitudes to the "Africans" and "Arabs" in Darfur. In the opposite direction, Darfuris will have opinions, presumptions, and anxieties about the intentions and biases of international forces in the area, so that the Marine commander will need to bear in mind what messages he sends by working with certain security forces or relief organizations.

The Marine who can balance these local cultural concerns with his tactical and operational imperatives is not guaranteed success—remember, the enemy always has a vote. However, he will be preparing his Marines to have more effect. The Marine presence will also be more sustainable, because it will be considered legitimate by sectors of the local population, whilst permitting Marines the situational awareness and clarity to take coercive measures when needed.

Case Study III: The Philippines – Kinship Politics and the Geographic-Religious Divide

As in Nigeria and Darfur, the physical geography and environment of the Philippines frame the historical and cultural features of the region. Stretching from Taiwan in the north to Malaysia and Indonesia in the south, the Philippines consist of over 7,100 islands connected by water, not land. As a starting point, therefore, it bears noting that there is no "Philippines" in the sense of a territorially-delimited nation after the Western model. Though Western powers from the sixteenth through twentieth century—the Spanish, Portuguese, Dutch, and Americans—may have drawn map lines and made alliances with local leaders amounting to a "Philippines," local diversity shows "Philippine nationess" to be an external construct, aiding local grandees of the Philippine state as well as international political and economic interests. As for the peoples of the Philippines, "state" and "nation" have historically been meaningful only as external intrusions into their daily and communal lives.

For centuries, due to their physical isolation from each other, most of the islands of the modern-day state of Philippines were politically, socially and economically independent. As a result, many of the islands developed their own unique identities, based upon environmental parameters, language diversity, economic relationships, and the political implications of local social structures. As just one example, today, it is estimated that over 100 languages are spoken in the Philippines, by a tremendous array of self-identifying ethnic groups. Not only do languages possess status, but when a person from a particular social grouping chooses to speak in a particular language, he/she is engaging in symbolic communication. Memories of functional independence, and diverse identities, influence local, regional, and national politics today, at times informing the way national leaders seek to manage crisis at the local level.

Migration and labor flows from both the west and northeast increased the cultural complexity of the region from the eighth through nineteenth century. This movement of people had tremendous and lasting influence on the economy, social structure, and belief systems of the islands. In particular, sea traders, first from India and China, and later from the Arab world, brought with them a variety of religions.

Religious importation began with Buddhism and Hinduism. Next came Islam, which effected a peaceful though substantial demographic shift in Mindanao, Basilan, Jolo, and the Sulu archipelago (and Borneo) as local populations converted over time. Each of these religions was incorporated into local languages and social as well as cultic practices. Local Islam, for example, emerged in syncretistic versions as people mixed indigenous "Little Traditions" with the major "Great Traditions."

As such, the physical geography of the country—an archipelago of separate islands—both causes and mirrors the ethnic, cultural, linguistic and religious diversity of the region. Environmental factors have led to challenges of transportation and communication among the islands. These challenges in turn have engendered a tendency to physical separatism and isolation based on hyper local social structures. Over the past five hundred years, these cultural dynamics have undermined successive efforts by the Spanish, the U.S. and the twentieth-century Philippine government to create a united centralized political system. In the Philippines, political structure reflects the environment, social structure, economy, and religion.

In 1565 the Spanish arrived, adding Catholicism to the complex belief system of the region. Spanish colonization and political control resulted in the conversion of many of the residents of the northern islands to Catholicism. The southern islands, however, which were culturally and religiously far more similar to the neighboring islands of Malaysia, remained staunchly Muslim. Unable to convert the southern islands, the Spanish colonists struggled for almost four centuries against a violent Muslim insurgency from their stronghold in Zamboanga. More broadly, the arrival of politically-supported Catholicism established a system of social stratification among foreign Christians, indigenous converts, pagans, and Muslims, which was layered over by existing linguistic, kin-based, and geographic strata.

It is during this time that the Muslims of the southern Philippines received the label "Moro." This is a pejorative term based on the word "Moors," which Spaniards had used for Arab and North

African Muslims whom they had ejected from the Iberian peninsula in the fourteenth century. Along with this being a nice illustration of "mirror-imaging," this label, and the ensuing centuries of Spanish-Muslim friction, also created new identities and the collective memories associated with them. Many Christianized Filipinos in the central and northern islands, particularly those who had intermarried with Spanish soldiers and administrators over several generations, came to view the Muslims as a violent social group, with this image captured in local folklore and folk drama.

Over time, Muslims, for their part, came to put a positive valuation on martial fierceness, though in their folklore, it was related more to brigandage and banditry, rather than to any ideological warfare for the sake of jihad. Part of this can be explained by the fact that concepts of "jihad" did not exist per se among Muslims in this region during this era—there is no direct ideological bridge between Muslim violence in the pre-twentieth-century era and that of today.

More fundamentally, and true to the pre-existing cultural dynamics, conversion to Islam did nothing to end the geographic, linguistic, and social disunity of the peoples of the southern islands. Put differently, Islam distinguished Muslims from others, but did little to unify them—a trend which has continued until today.

The arrival of the Spanish also caused a shift in both residence patterns and economy for the people in the Philippines islands, particularly in the central and northern islands. Prior to Spanish colonization, most people resided in simple bamboo huts scattered along the ocean or rivers. Some earned a living by fishing, but most were either tenant farmers or ran their own small independent farms. With the Spanish came Catholic missions and the appearance of mission town complexes built around a central plaza. Rural inhabitants were encouraged to move to the towns which offered such attractions as the availability of a greater variety of local and imported goods, free Catholic school for children, and urban amenities. As geographical and physical space was altered, new patterns of movement and socialization emerged. As we will see soon, these new spatial and geographical patterns would serve to strengthen a local politics based on elite families.

The Spanish also introduced a variety of new crops intended for export, including sugar cane, tobacco and abaca—a strong fiber used to make twine. The new export crop economy decreased previously independent small farmers' abilities to engage in successful subsistence agriculture. As in other areas, the new economy was accompanied by the development of large haciendas supported by a tenant farmer population. This new economy concentrated wealth in the hands of an elite group of families. These family groupings emerged as the nodes of formal political power in local government positions. Locally, and later, nationally, these families formed a loose oligarchy, or governing council. Economic power thus became intimately linked to political power in a hierarchical social structure based on kinship.

Complicating the cultural picture was the development of a parallel religious power structure: the Catholic Church. Missions owned large tracts of land and played a major role in the agricultural economy of the colony. Although priests could not govern, they held great power over their parishes. As a result, a parallel power system operated side by side with the formal government, and the Catholic clergy often acted as informal leaders in many regions.

Initially only Spaniards could be ordained, concentrating religious power in the hands of the colonizers. Since there were far fewer Spanish priests than available positions, many regions—especially the more rural areas and remote islands—went without formal religious leadership or training. As a result, Catholicism in the more remote and rural regions was a complex mix of local superstitions and traditions interspersed with fragments of Catholic teaching. Of course, because traditional Islam has no clerical hierarchy and because of the communal disunity of the region's Muslims, no parallel religious power structure emerged among the Muslims.

After the 1898 U.S. victory in the Spanish-American War, control over the Philippines was ceded to the U.S. Although never effectively ruled by the Spanish, the southern islands, including the major island of Mindanao, were included in the cession. In contrast to British, French and Dutch colonial regimes which focused on centralizing political authority and power, U.S. policy in the

Philippines supported the formation of a decentralized state. This reflected American notions of governance, including a strong bias for local autonomy placing limits on the power of a central authority.

This American bias unwittingly supported the indigenous tendency towards decentralized local bosses; a political system run by powerful local kin groups or oligarchies; and private police and militaries whose allegiance was to regional, rather than national, powers. Such a system was highly open to corruption and political abuse, since no centralized political authority had oversight or control over the economic, social, or political activities of the local bosses.

With the majority of power and wealth concentrated within an elite group of families, clear class differences continued to emerge. These class differences became even more severe as the economy shifted from a primarily agricultural base to an industrial one. Factories sprung up to process the cane sugar prior to export; abaca and other plants were processed in textile factories for the growing world market; and rice was processed and cleaned prior to export from growing industrial cities such as Manila and Luzon. Long involved in Philippine trade, Chinese and Indian merchants began to settle in large numbers in these cities, creating a sizeable urban immigrant population.

The growing economic disparities between the wealthy elite and the poor tenant farmers and factory workers led to the development of several socialist and Communist parties and movements from the 1930s to 1950s. The most famous of these was the HUKBALAHAP (Hukbo ng Bayan Laban sa Hapon or National Anti-Japanese Army), nicknamed the Huks. Allegedly organized to fight the Japanese during World War II, this group was actually a political arm of the Philippine Communist Party. Challenging the class and kinship systems that had controlled most of the political and economic power of the country for several centuries, these Communist groups gained much popular support among the poor and economically marginalized groups of the country. Among Muslims, however, the appeal of communist associations has been less, not only because of Marxist atheism, but also because these groups emerged from within the Christian Filipino community.

U.S. policies not only supported and entrenched existing kin and class structures, but they also increased religious polarization in the country, which has manifested itself in several ways. First, unlike the Spanish, the U.S. administrators possessed something like a language policy, in that they encouraged the spread of English, which ultimately became an official language, along with Tagalog. The public education system and press in the northern islands promoted English, allowing cross-community bonds to develop, particularly as rural to urban migration increased in the era of industrialization.

The Muslims, however, did not benefit to nearly the same extent from English language education or media. Rather, different Muslim communities, while increasingly learning Tagalog, persisted in the use of their highly localized regional languages, such as Yakan, Samalan, and Tausug. Identity divisions followed the geographic-linguistic fault lines, creating yet another barrier to both Muslim unity in the south and an effective Muslim political voice in Manila. While Tagalog and English are today shared languages among Muslim leaders in Mindanao and Sulu, Tagalog has come to be closely associated with the Christian Filipinos. Therefore English is a less loaded common language, for foreigners as well.

Second, in order to stabilize the southern Muslim islands, the U.S. encouraged settlement by non-Muslims from the north. This policy was supported by a series of Public Lands Acts, which removed many of the ancestral lands from local Moros and redistributed them to incoming Christian settlers. Increasing the sense of disenfranchisement, the government encouraged the establishment in the south of large corporations, whose primary purpose was to create factory products for export. As a result, the government shifted the economic base of land and industry away from the Moros and into the hands of outsiders. Economic and political structures thus ensured the crystallization of a heretofore latent Muslim-Christian fault line among indigenous people.

In 1946 the Philippines became independent. The new government faced a daunting legacy: an enormous archipelago composed of almost one hundred diverse cultural and linguistic groups. For both Muslims and Christians, governance was concentrated in the hands of specific elite families that also controlled most of the land and business activities of their regions. Either as mayors, landowners, governors, or police chiefs, hereditary local leaders represented the interests of economically advantaged families, who lackadaisically enforced the central government's writ while creating webs of patron-client relationships. Into the 1960s and 1970s, both to protect family interests and as a hedge against potential Muslim-Christian violence, local leaders relied on militias to support their political disputes.

Local disputes were often expressed through the idiom of the two main national-level political parties. Still, disputes reflected regional factional interests, and bore no relation to the ideological coloring of those parties. Attempts in the 1960s by groups such as the Muslim Independence Movement to transcend the hyper-local orientation did not succeed in uniting Muslims, but did keep embers of Muslim insurgency smoldering.

The lack of effective and impartial centralized authority provided much opportunity for corruption at both the local and national level. Despite efforts by various regimes to challenge the economic and political bases of power in the country, the Philippines remained hostage to *politica de familia*, or kinship politics, throughout the 1970s and 1980s. While Ferdinand Marcos abrogated the constitution and ruled by decree from 1972 in the name of increasing the efficacy of the state and rooting out corruption, he did not in fact do this. Instead, his family replaced previous kin groups as holders of political and economic power.

National-level representation of Muslim Filipinos did increase during this period, along with greater levels of education and attendance of the national university in Manila. However, the Marcos years further marginalized the Muslim Filipinos of the south, who were also disadvantaged through the new labor arrangements of an export-oriented agricultural and industrial economy. The regime's moves to extend the army's control of the south also caused a growth in violence. Central Mindanao and Sulu were targeted by successive campaigns, during which Jolo was razed and Basilan hit very hard. These anti-insurgent moves set in train refugee flows to more central parts of the country and Manila, creating a new social group that would become important in the 1990s and beyond.

While bringing violence and a visible state coercive presence, these efforts did not hinder lawlessness or the proliferation of non-state armed agents. By the end of the 1970s, therefore, the Marcos regime agreed in principle to regional autonomy in the southern regions, but groups in the Muslim south opposed the agreement. This disagreement was expressed in Islamic and political terms, yet was in fact related to local political feuding and ethnic differences among the area's Muslims.

The 1986 "EDSA Revolution"[3] that unseated Marcos in favor of Corazon Aquino also did not break this pattern of *politica de familia* either at the local or at the national level. Descendants of the elite families who came into the first post-1898 legislature continue to dominate in parliament, though now they have brought celebrity actors, singers, and athletes into their families to shore up their political power. This kind of politics, whereby party affiliation became meaningless, limped along into 2001, when the president of the day sought to use his patron-client connections with parliamentarians to defeat a looming impeachment. Public backlash emerged again, resulting in the "2001 EDSA Revolution." The nature of politics, based on a specific kin-based social structure that emerged from economic relationships, has not changed since.

The 1980s-1990s did witness evolution in the geographic, social, and organizational bases of Muslim insurgency in the south. In the 1970s, a series of Muslim movements emerged in the region, often breaking away from existing groups based on regional, linguistic, or ethnic differences. Some of these groups, in the 1980s, were led by younger, more modern-educated leaders who sought to challenge not only Christian Filipino rule, but the traditional Muslim elites. And though the heretofore dominant group, Moro National Liberation Front (MNLF; its armed faction is the Bangsa

Moro Army) had agreed by the late 1980s and early 1990s to negotiate for an autonomous region, other groups demanded independence instead.

In 1995, an attack on the Christian town of Ipil on Mindanao heralded the emergence of a new group, much more militant in nature, that has become connected to international Islamist terrorism. Known as the Abu Sayyaf Group, their social base is unlike any of the existing Muslim groups. Emerging from the Christian city of Zamboanga, it could call upon the large numbers of displaced young Muslims whose homes were destroyed during the state's campaigns of the 1980s and 1990s. These refugees had been removed from both their ethnic and linguistic homes, and suffered from unemployment and a certain amount of social exclusion in the urban area.

The MNLF, led by secular Muslims with university educations, was not of the same social stratum as these younger refugees, and did not speak in the same political idiom. Other groups, such as the Moro Islamic Liberation Front, represent the established social elites, with no connection to this new demographic group whom both the state and traditional Moro society have excluded.

By contrast, the Abu Sayyaf Group's use of militant, non-compromising Islamist rhetoric, and the more puritanical ideological leanings of its first leader, Abubakar Janjalani (since killed by Philippine forces), have been able to attract these Filipinos. This attraction has been able to transcend enduring ethno-linguistic divisions among Muslims in the area, attracting Magindanao, Maranao, and Yakan and Tausug people—the latter having long been rival communities.

Abu Sayyaf's particular appeal is possible not only because economic and political conditions created new social formations, but also because Abu Sayyaf is a break from existing class- and ethnicity-based movements among the Muslims. Furthermore, it can tap into the "outlaw" motif that has long been a part of islanders' identities.

Since 2001, the southern Philippines has been part of Operation Enduring Freedom. U.S. special operations forces and Marines have trained and logistically supported Philippine forces in the south, who have racked up notable successes in arresting or killing insurgents. Perhaps of more enduring counterinsurgency value, U.S. civil affairs assistance has permitted the Filipino forces to build schools, wells, and markets, often after they have been destroyed by terrorists. This has earned the local military some credit in the eyes of the non-combatant southern Muslims.

As we have seen here, long-term solutions to both political instability and southern insurgency are contingent on more enduring issues. The problems are not simply related to corruption or an inefficient democratic political system—and are not, in the main, a function of "fundamentalist" or "jihadist" ideology. Rather, challenges in the Philippines result from the unique interactions among the Five Operational Cultural Dimensions presented in this book.

In this perspective, the geographical conditions of island isolation reinforce local power and frustrates the national government's capacity to exercise leadership at the local level. Social structures of kinship and class concentrate economic and political power among a select set of elite families, while multiple linguistic and ethnic group memberships are used to reinforce concentrations of power and demarcate excluded groups. Not only have geographic, linguistic and historical factors resulted in religious differences between the northern and southern islands, but these same kinds of differences among the Muslims of the south frustrate attempts at Muslim unity.

Finally, the way the Operational Culture Dimensions of environment, economy, social structure, and political structure interact with each other has resulted in political and economic marginalization of the south, and the organized expression of resistance to these conditions—the MNLF, MILF, and Abu Sayyaf Group have relied on evolving social groups for a basis of support. Today, identities associated either with group memberships; themes in local folklore; memories of recent and "age-old" history; or ideas of privilege vs. dispossession have ensured that only sustained engagement of

Filipino leaders at the national and local level, supported by the U.S., will lead to a long-term solution in an area of vital concern to American interests.[4]

Likely Marine missions in the Philippines include mobile training teams focused on a particular Philippine branch of service or element; battalion-size exercises with the Armed Forces of the Philippines; civil affairs support; or humanitarian affairs and disaster relief operations. The likelihood of large scale stability or counter-insurgent operations in which U.S. forces are a lead element is low at this time. Still, there are particular cultural considerations to bear in mind, no matter what kind of deployment.

The first is very practical. Given the large number of hyper-local languages in the southern islands of the Philippines, and the ethnic, religious, and political overtones associated with several of them, a commander will need to determine what languages to focus pre-deployment training on, and what languages to ensure interpreters know. Though Tagalog will be tempting, use of this language may associate U.S. forces too closely with the Christian political leadership in Manila. Therefore, when communicating with indigenous civilians, English may in fact serve best.

Second, in the context of humanitarian affairs, disaster relief, or civil affairs projects, a commander will interact with local political and economic leaders. These leaders' positions will all be contested, and as such their decisions will need to integrate hyper-local, kin-based considerations. As such, American assistance will become part of local power-plays. Yet, absent the integration of local leaders, be they elite, uneducated, Islamist, or even outlaws, U.S. initiatives may lack legitimacy, the essential factor permitting enduring success.

Either way, any interaction with local Filipinos, in the context of contracting, humanitarian support, or mentorship of governance, will put Marines in a position to witness corruption, misuse of funds, and misuse of authority. In local terms, this corruption will be necessary for leaders—or would-be leaders—to create, maintain, and expand partially kin-based patron-client networks.

Third, Marines may encounter elevated levels of crime. It will be important to distinguish between "criminal" crime and insurgent activity. Particularly as motifs in local folklore and self-identity put a positive valuation on "wanted" people, "outlaws," "bandits," etc., commanders at every level will need to consider how American norms differ from local views on this issue. This will assist to avoid tarring all organized criminal activity as pro-Abu Sayyaf.

Of course, since the effect of such activity may be to reduce confidence in state authorities to the point that insurgent groups appear as better alternatives, Marines will need to encourage local forces to engage in the kinds of counter-crime operations that increase security while earning enough local credit to diminish local support for insurgency. In the same vein, information operations planners will need to pay particularly close attention to these local folk motifs, to ensure that IO, executed either by U.S. personnel or by indigenous forces, aligns with local motifs.

Fourth, a commander will need to consider who his forces are associated with in the eyes of multiple constituencies in the region. Twentieth-century American colonialism benefited the north more than the south, and brought with it more exploitative economic relations in the south. Local insurgents or contending kin groups might seek to activate this memory among Filipinos, tarring the U.S. forces as neo-colonizers. Close attention to Philippine state sovereignty will thus be essential, though it may result in Philippine actions that are inefficient or counter-productive from a counterinsurgency perspective.

Conversely, ongoing U.S. economic interests in the region, and a track-record of supporting political leaders who have not helped the south, might aggravate a memory of American collusion with local oppressors. Too close an association with the Philippine government and security forces may decrease local inclination to work with the U.S.. This all argues for a hyper-local focus with respect to points of contact and cooperation—though a heretofore intrusive state may not prefer this focus on local people at the expense of Manila.

A fifth training, planning, and operational consideration has more to do with the commander's Marines than it does with the Filipinos. It is likely that many of the deploying troops will have served in Iraq or Afghanistan. They will have worked with, or fought against, Muslims. They will have learned about Islamist extremism and encountered its propaganda. It will be tempting for these Marines to view the culture of their current deployment in terms of the Muslim culture of previous deployments. Abu Sayyaf may thus appear entirely through the prism of al-Qaida, Abu Musab al-Zarqawi, the Mahdi Army, Taliban, or other violent groups from Arab lands or South Asia. It will be extremely important to focus Marines at all levels away from these previous experiences, and focus them on the particular ethnic and linguistic dynamics in the Philippines.

Likewise, rather than "seeing mooj," Marines at all levels will need to learn about the specific social strata which are likely to be excluded economically, socially, and politically, so that indigenous attraction to Abu Sayyaf, as members, readers of propaganda, or in other ways, will be seen for what it is, and not layered with experiences in OIF or Afghanistan.

By the same token, in order to have the proper mental orientation to establish rapport with Filipino Muslims, commanders will need to ensure that their troops do not perceive a common cause with Christians based on shared religion. After all, a tradition of syncretism has led to a Philippine expression of Christianity different from that in the U.S.

This impartiality will not only be important to preserve, but potentially difficult to do. As Marines train with, support, and build relationships with Philippine police and military, they will be interacting predominantly with a Christian force. These soldiers' families are either from the north or, worse, from contested regions where Muslim-Christian friction has resulted in bloodshed. These soldiers also may have experiences of fighting "Muslim terrorists" during their or their relatives' military service. It is even more likely that there are memories in their families of "age-old" conflicts with Muslims. As interlocutors or indigenous sources about south

Philippine culture therefore, Marines' Philippine army brethren would provide a biased picture.

Conclusion

The above case studies show that the Five Operational Culture Dimensions—physical environment, economic system, social structure, political structure, and beliefs and symbols—do in fact illustrate dynamics at play in areas to which Marines might deploy. Without surveying an area of operations in terms of these dimensions, and searching for the linkages among them, a Marine will not have an integrated cultural picture of the operating environment, and any cultural knowledge he does glean will thus remain incomplete and misleading as to cause, effect, and proper measures to take.

Notes

[1] Jeffrey Tayler, *Angry Wind: Through Muslim Black Africa by Truck, Bus, Boat, and Camel* (Boston, MA: Houghton Mifflin, 2005); Idem., "Worse Than Iraq?" *The Atlantic Monthly* (April 2006); Karl Maier, *This House Has Fallen: Nigeria in Crisis* (Boulder, CO: Westview Press, 2003); Tom O'Neill "Curse of the Black Gold: Hope and Betrayal in the Niger Delta," *National Geographic* (February 2007); Daniel Jordan Smith, *A Culture of Corruption: Everyday Deception and Popular Discontent in Nigeria* (Princeton, NJ: Princeton University Press, 2006); Rotimi T. Suberu, *Federalism and Ethnic Conflict in Nigeria* (Washington, D.C.: U.S. Institute of Peace Press, 2001).

[2] Julie Flint and Alex De Waal, *Darfur: A Short History of a Long War* (London: Zed Books, 2006); Stephan Faris, "The Real Roots of Darfur," *The Atlantic Monthly* (April 2007); Mahmood Mamdani, "The Politics of Naming: Genocide, Civil War, Insurgency," *London Review of Books* 29:5 (March 2007); Ann Mosely Lesch, *The Sudan: Contested National Identities* (Bloomington, IN: Indiana University Press, 1999); Douglas Hamilton Johnson, *The Root Causes of Sudan's Civil Wars* (Bloomington, IN: Indiana University Press, 2003); J. Milard Burr and Robert O. Collins, *Darfur: The Long Road to Disaster* (Princeton, NJ: Markus Wiener Publishers, 2006).

[3] "EDSA" stands for Epifanio de los Santos Avenue, a main thoroughfare in Manila where demonstrations occurred.

[4] Jose S. Arcilla, *An Introduction to Philippine History* (Manila: Ateneo de Manila University Press, 2003); Paul D. Hutchcroft, "Colonial Masters, National Politicos and Provincial Lords: Authority and Local Autonomy in the American

Philippines, 1900-1913," *The Journal of Asian Studies* 59:2 (2000): 277-306; Syed S. Islam, "The Islamic Independence Movements in Patani of Thailand and Mindanao of the Philippines," *Asian Survey* 38:5 (1998): 441-456; Charles O. Frake, "Abu Sayyaf: Displays of Violence and the Proliferation of Contested Identities among Philippine Muslims," *American Anthropologist* 100:1 (March 1998), 41-54; Sheila Coronel, Yvonne Chua, Luz Rimban, and Booma Cruz, "The Rulemakers: How the Wealthy and Well-Born Dominate Congress," *Philippine Centre for Investigative Journalism*, 2004; James Hookway, "U.S. War on Terror Shows Promise in the Philippines: Local Officers Lead, with American Aid," *Wall Street Journal* 18 June 2007, A1; Idem., "Terror Cells Team Up to Survive in the Philippines," *Wall Street Journal* 22 Jan 2007, A8; Simon Montlake, "Where the U.S. is Helping to Make Gains Against Terrorism," *Christian Science Monitor* 15 February 2007, 1; Rommel C. Banlaoi, "Maritime Terrorism in Southeast Asia," *Naval War College Review* 58:4 (Autumn 2005); Mark Turner, "The Management of Violence in a Conflict Organization: The Case of the Abu Sayyaf," *Public Organization Review: A Global Journal* 3 (2003), 387-401; Gary Hawes, "Theories of Peasant Revolution: A Critique from the Philippines," *World Politics* 42:2 (1990); 261-298; Mel C. Labrador, "The Philippines in 2000: In Search of a Silver Lining," *Asian Survey* 41:1 (Jan-Feb 2001), 221-229.

Chapter 9

Operational Culture Learning in Training and Education

Part I of *Operational Culture for the Warfighter* focused on debates and definitions. Part II went on to examine the core Operational Culture Dimensions for planning and execution in multiple cultural environments. In Part III, Chapter Eight presented case studies highlighting the interaction of the Five Operational Culture Dimensions, as well as their implications for potential Marine deployments. The present chapter is both conceptual and applied. It seeks to situate the discussion of culture learning within the twin realms of pedagogical science and military training and education.

We begin here by presenting Bloom's Taxonomy of learning, a widely accepted taxonomy applicable to both academic education and skills training. We do this to ensure that the discussion of culture learning adheres to rigorous intellectual standards, and avoids the pitfalls of some recent academic and military writing on the subject. We then move on to illustrate how and why this matters to the particular field of learning about cultures for the sake of military operations.

On this basis, Chapter Nine uses Bloom's Taxonomy as the vehicle to analyze the dominant modes of learning in the military today. These include pre-deployment training programs (PTP) and professional military education (PME) across the phases of an enlisted or commissioned career. In each case, we relate subject matter and learning goals to different elements of Bloom's Taxonomy. We conclude the chapter by considering the kinds of subject matter expertise and experience necessary for educators in the realm of operational culture.

Initial Issues for Consideration

Before we can move into the main substantive parts of this chapter, there are certain **over-arching considerations** to bear in mind while reading the chapter as a whole. Some we deal with directly; others are intended as themes for consideration throughout. **First**, as has been suggested throughout this book by use of terms such as "study," "examine," "detect," "investigate," etc., operational culture capability emerges mostly from learning, be it as an individual Marine, as an operational unit, or in the formal schools of the Marine Corps.

In fact, learning about individual cultures, and grasping the tools and skills to use in foreign cultures generally, is a *shared* responsibility. Every individual Marine in a leadership and mentoring role must take his or her own initiative to learn about cultures and cultural concepts, based both on the prism of their own operational experience as well as on academic and conceptual learning. The individual Marine also has the responsibility to facilitate cultural learning among subordinates, either as individuals or through the schools of the Marine Corps.

While Marine Corps Training and Education Command (TECOM) develops curricula and standards for use in schools and the fleet, commanders and individual Marines themselves must ultimately participate actively in education and training, and pursue self-study as appropriate to mission and rank. Again, in addition to ensuring a substantive grasp among their Marines, active leadership participation in culture learning reinforces the idea among subordinates that every Marine is a "culture operator," just as "every Marine is a rifleman." This is because in any organization, the greatest change mechanism is socialization, and in the Marine Corps, that socialization mechanism is training and education, both formal and informal.

Second, while culture learning should target specific regions when appropriate to mission or educational goals, more fundamental and capability-building learning focuses on the basic concepts and skills with which to "read" any culture—in short, it is about the Operational Culture Dimensions in this book. Thus, while learning information about an AO will help for a specific deployment, a more conceptually-focused approach to Operational Culture learning will serve the Marine better throughout his or her career.

Most basically, this is because conceptual knowledge can literally "travel anywhere," and can be applied to diverse environments. Furthermore, as Marine Corps doctrine reminds us, one of the greatest benefits of operational preparation and planning is the process for developing "a common understanding of the nature of the problem," through "a disciplined framework for approaching problems… without restricting judgment and creativity."[1] It is through integrated Operational Culture learning spanning PTP, PME, and career-long learning, and based on the dimensions discussed in this book, that Marines at all levels will develop that "common understanding" of what culture means in the battlespace, in the form of a "disciplined framework" with respect to culture. Finally, if pursued in this fashion, training and education will create a Marine who understands culture as a core combat competency, regardless of area of operations or any given operation's location on the kinetic spectrum.

Third, learning effectively about cultures in order to maintain fidelity both to the military mission *and* the phenomenon of "culture" itself means rethinking the military approach to training and education—*for culture learning*. That is not to say that existing models of military learning—in particular "The Systems Approach to Training"—are not effective with respect to training and education for traditional military functions and procedures.   Rather, as we saw in Part I, due to the nature of culture as internally varied; fluid and dynamic; and dependent on other factors in explaining individual behavior, it is somewhat different from traditional fields of military learning. Therefore, enduring conceptual and functional learning about culture requires a more hybrid model for conducting and evaluating learning. We will return to this topic in subsequent sections of this chapter.[2]

Learning Domains

In order to consider operational culture learning in training and education, we begin with organizing ideas of learning into a proper framework. To organize learning concepts, it helps to think from the perspective of Bloom's Taxonomy of Learning.[3] This taxonomy has several benefits in a discussion of culture learning. **First**, it has been around for quite a while, and is a commonly accepted method. Bloom's Taxonomy has thus established the categories for much of the subsequent literature on learning taxonomies in the fields of education and psychology. **Second**, this taxonomy also works well with the Marine Corps' Systems Approach to Training, though it also proves useful for hybrid fields of learning, such as culture.

Third, this taxonomy renders more systematic and disciplined our thinking. This is important because recent attempts to fit culture into a framework of learning for military applications have used words—such as "consideration," "awareness," "knowledge," "understanding," and "competence"—to describe knowledge hierarchies. These terms and hierarchies are then assigned to ranks across the force.[4]

This all sounds scientific, and with the help of an accompanying pyramid diagram and rank correlations, this method helps us to visually pattern culture learning in a way that is easily digestible—particularly to military readers. However, these terms are not linked to any meaningful and accepted taxonomy of learning, so that their definitions are arbitrary, as are their rank correlations. We thus choose to use Bloom's Taxonomy as a guide to our discussion.

According to Bloom's Taxonomy, learning is divided into three chief domains termed cognitive, psycho-motor, and affective. To define these terms briefly:

Cognitive Skills: The mental processes for gaining knowledge and comprehension. These skills involve thinking, recalling, and judging. At higher levels, cognition assists imagination and planning.

Psycho-Motor Skills: These behavioral skills involve movement of the body in order to perform actions based upon input from the brain.

Affective Skills: These skills are based on feelings and emotions that are subjective, and emerge from one's own mind and attitudes towards observable events and data. Affective skills involve how a person relates to and interacts with what they learn, and how they come to own it both intellectually and psycho-emotionally.

As a shorthand:

Cognitive domain	=	thinking domain
Psycho-**Motor** domain	=	doing domain
Affective domain	=	feeling domain

Each of these domains yields several ascending skills, developed according to one's native capabilities and exposure to material.

Figure 9.1: Bloom's Taxonomy of Learning

Domain	Ascending Levels of Learning					
	1	2	3	4	5	6
Cognitive	Knowledge	Comprehension	Application	Analysis	Synthesis	Evaluation
Psycho Motor	Imitation	Manipulation	Precision	Articulation	Naturalization	
Affective	Receiving	Responding	Valuing	Organization	Values Complex	

This taxonomy is useful not only because of the characteristics of each descriptor across the domains, but also because it posits a hierarchy of learning—keyed to the learning activity itself, and not rank, age, or other arbitrary criteria. We will now describe each domain in detail, beginning with an explanation of the meaning of each descriptor in the domains.

Table 9.1: Cognitive Domain Descriptors and Characteristics

Cognitive Domain of Learning		
Descriptor		**Characteristics**
1	Knowledge	Remembering previously learned material by recalling information
2	Comprehension	Grasping meaning of material through explaining or summarizing, and estimating consequences of actions
3	Application	Using learned material in new/concrete situations, by applying rules, methods, concepts, and principles
4	Analysis	Breaking down material into its component parts to grasp its organizational structure and the relationship among its parts
5	Synthesis	Putting parts together to form a new whole, as a plan or classification system
6	Evaluation	Judging the value of material according to internally-generated or externally-provided criteria

Table 9.2: Psycho-Motor Domain Descriptors and Characteristics

Psycho-Motor Domain of Learning		
Descriptor		**Characteristics**
1	Imitation	Repeating an act that has been demonstrated or explained, with trial and error until achieving appropriate response
2	Manipulation	Practicing a particular skill until it becomes habitual and can be performed with some proficiency, but not certainty
3	Precision	Proficiency through smooth performance without hesitation, which can be complex
4	Articulation	Precision so well developed that movement patterns can be modified to fit special requirements or problems
5	Naturalization	Response is automatic, permitting new motor acts or ways of manipulating materials out of understandings and skills developed

Table 9.3: Affective Domain Descriptors and Characteristics

Affective Domain of Learning		
Descriptor		**Characteristics**
1	Receiving	Willingness to receive learning through awareness and selected attention
2	Responding	Active participation with sufficient commitment to a subject or activity to seek it
3	Valuing	Seeing worth /value in subject, motivated by commitment to underlying and clearly identifiable value guiding behavior
4	Organization	Bringing together values, and beginning to build an internally consistent value system related to already held values, in a dynamic equilibrium dependent upon events at specific points in time
5	Value Complex	Internalization of values in the individual's value hierarchy, creating a characteristic "life style" that is pervasive

The above description of Bloom's Taxonomy of Learning lays out the basic principles and levels of learning across different skill domains. It is thus suited to learning about culture for military applications, for several reasons. **First**, because use of culture in a military context requires both "hard" and "soft" skills, effective culture operations require learning across all Bloom's domains: Knowing "stuff" is not enough; neither is simply being able to perform tasks, or feeling some sort of disorganized, mission-unrelated empathy for foreign peoples. Likewise, operational culture learning needs to occur across these domains simultaneously, and in a graduated fashion throughout one's career.

Second, learning in these domains is characterized by certain skills and mental approaches, some of which—particularly the cognitive and psycho-motor skills—are observable. **Third**, different learning contexts—professional military education, pre-deployment training, or individual regional expertise—require particular focus on specific domains or skills.

Fourth, and of tremendous significance, Bloom's Taxonomy includes the affective domain of learning. This is extremely important to culture learning. In the military, training and education explicitly embrace and value a learning model based on the cognitive—"what do you know"—and psycho-motor—"what can you do"—domains. However, there is a third kind of intelligence less associated with traditional military competencies, but which is essential for the kinds of missions Marines will undertake today and in the future. This trait, developed through learning in the affective domain, is Social Intelligence.

Resulting from experiences, learning, and intuition, **Social Intelligence** is not a sub-discipline of intelligence as an MOS, such as air intelligence, signals intelligence, etc. Rather, it is used to connote certain capabilities not usually taught in school or trained through task performance, but which develop through a cascading process of combining experience, reflection, and self-knowledge. Social intelligence refers to the intellectual ability to understand oneself and others in social interactions.[5] Others have described this ability as "emotional intelligence"—the ability to understand and regulate one's own emotional/mental state and reactions to stressful events, and also to better perceive the emotional state of others with whom one interacts.[6]

Social and emotional intelligence include certain traits—some trainable, and some gained through experiential learning—quite significant for operational culture learning and for extended operations among foreign peoples and with foreign partners:

1) Emotional Self-Awareness　　2) Independence
3) Inter-personal relationships　　4) Empathy
5) Stress Tolerance　　6) Impulse Control
7) Flexibility　　8) Problem Solving

Further, educators and researchers emerging from psychology, international business, and learning sciences relate concepts of social intelligence to the ability to function in other cultures. These writers refer to this ability as "cultural intelligence."[7] Since the U.S. military also uses this term, but with the connotation of information and intelligence about foreign cultures, we will here refer to these concepts as "cross cultural effectiveness." Cross cultural effectiveness involves a cluster of cognitive and performative skills:

Perceptual Skills:	Open-mindedness, tolerance of uncertainty, non-judgmentalness
Relational Skills:	Flexibility, sociability, empathy
Adaptive Skills:	Ability to demonstrate behaviors from a repertoire gained through experience of multiple cultures, as well as the particular one at a given time

Any Marine who has mentored Iraqi, Afghan, or other countries' security personnel—or for that matter recruits and junior Marines—will look at the above lists and see qualities that have proven necessary for operational success. They can thus attest to the importance of social intelligence. In effect, to successfully integrate culture learning in operations, a Marine should combine social intelligence with prudent evaluation of tactical considerations.[8]

In order for individual Marines to gain the necessary social and emotional intelligence to be culturally successful in the cognitive and psycho-motor domains—as well as when deployed—their socialization as recruits, training in units, experience at schools, and mentorship of fellow Marines should incorporate culture into the affective domain of what it means to be a Marine. At the same time, these experiences must incorporate the affective domain into culture learning.

The affective domain of learning, as well as social and emotional intelligence, are also essential to an additional capability which is also distinct from Operational Culture: cross-cultural competence. Recent efforts in the Military to measure the ability to function successfully in other cultures have resulted in a number of studies and programs on this topic.[9] Educationally, it is important to understand that cross-cultural competence focuses on an "intelligence" (as in knowledge ability) and its measurability in terms of testable scales. It is a psychological concept focusing on the Marine or service member being deployed. Cross cultural competence scales are popular with the military because a) they are measurable; b) one might be able to teach and train to them; and c) a good scale should provide some predictive ability about who will and will not adjust well in another culture.

It is important here to note that cross-cultural competence is not a synonym for Operational Culture, nor is it an alternative way of understanding culture as a concept, or culture groups around the world. Operational Culture, as described in this book, is an anthropological approach which focuses not on the *Marine* going overseas but on the *people* s/he will meet, and the *environments* in which s/he will meet them. Put differently, cross cultural competency focuses on the adaptability of a person to a culture, while Operational Culture focuses on the ways to understand, plan for, and operate in those cultures to which Marines may have to adapt. As a set of concepts that can be applied in planning and operating, Operational Culture can help the Marine evaluate conditions in foreign areas in terms of commander's intent and desired effects. Using these concepts, a Marine can develop effective courses of action based on cultural factors affecting planning and operating.

Operational Culture and cross-cultural competency are neither opposing approaches to culture nor divergent capabilities. Education and training emphasizing affective-domain learning will more likely cultivate both cross-cultural competence *and* an inclination to think and operate in terms of the Five Operational Culture Dimensions. Absent this cross-cultural competence, the likelihood of possessing the social and emotional intelligence to function in foreign environments will be lower, and the affective component of Operational Culture Learning will be much diminished. Still, without learning about foreign cultures in terms of the Five Operational Culture Dimensions, one is unlikely to have the skills necessary to plan and execute foreign area operations.

Any framework for culture learning must address the affective domain. Programs and measures which evaluate and develop Marines' cross-cultural competence skills will preserve the capability of Marines to be America's high-intensity force of unmatched lethality and speed, while granting them the social intelligence to plan and function successfully in foreign cultures.

Example: *The Affective Component Makes an Effective Marine*

If we consider the culture group known as "Marines" from the perspective of the Operational Culture Dimension of beliefs and icons, we see that they are unique. As a high-intensity force of unmatched lethality and speed, "Magnificent Bastards," "Ugly Angels," and "Devil Dogs" know how to "get some" and have proven adept at it, and attitudinally predisposed to do so, from Belleau Wood to the "Frozen Chosin" to Fallujah.[10]

This unmatched lethality, destructive capability, and speed do not solely or mostly come from a combination of gear, training, and doctrine. Rather, one could argue that most important in making a Marine is the *attitudinal predisposition* to be lethal, destructive, and fast—and better at it than anyone else. That is why men and women choose the Corps. Is is also what recruits are socialized not only to *know*, but *to believe and feel as "who I am."* In brief, Marines are good at what they do because they really want to *do* it, and want to *become* it, even at the expense of other elements of life—Marines incorporate "Marineness" into the affective dimension of their conduct, explicit learning, and implicit socialization, and this guarantees cognitive and psycho-motor success in the FMF.

Example: *The Affective Component Makes an Effective Marine;* continued.

Furthermore, through screening officer candidates and recruits, and while at Boot Camp and OCS, Marine learning implicitly emphasizes experiences meant to cultivate traits of emotional intelligence. Traits of emotional intelligence that receive focus in Marine entry level training include independence; inter-personal relationships; stress tolerance; impulse control; flexibility; and problem solving. Whether it be the Crucible or the Small Unit Leader Evaluation (SULE II), Marine training in fact puts a high value on emotional intelligence along with affective learning—though no Marine would admit to emotions!

Samples of Operational Culture Learning Across the Domains

Before we move on to consider different kinds of learning settings from the standpoint of Bloom's domains, we first need to give these domains some context with respect to both culture and military operations. Therefore, we will review Bloom's descriptors here, including concrete examples of what they would entail for Marines learning about a specific cultural environment, in this case a Muslim one in the Middle East.

As these tables below show, it is indeed possible to develop concrete training and education about cultures in a graduated fashion across the domains of learning. Within each domain of Bloom's Taxonomy, not only does the learning build upon preceding learning outcomes, but each stage requires a higher level of capability and understanding. These higher levels of capability in the cognitive and psycho-motor domain, in turn, demand equal progress through the rungs of the affective domain. On this basis, we can now proceed to a discussion of Operational Culture Learning in both pre-deployment training and professional military education.

Table 9.4: Cognitive Learning in Cultural Context

Cognitive Domain of Learning	
Descriptor	**Characteristics**
Knowledge	Enumerating numbers of prayer times in Islam
	Remembering the tribal map of the AO
Comprehension	Explaining main activities in Muslim prayer
	Estimating implications of detaining family heads
Application	Applying understanding of Muslim prayer and mosque architecture to determine roles of Marines and indigenous security forces in entering a mosque, dialoguing with leaders, searching men and women
	Determining which indigenous participants in a meeting are leaders, and whose authority is based on religion, family leadership, coercive power, etc.
Analysis	Grasping interrelationship of religious membership and ethnicity, and how relates to politics, crime, and challenges to stability
	Deconstructing AO's social structure in terms of age, gender, class, kin, ethnicity
Synthesis	Assessing AO in terms of mission objectives, bringing together considerations of environment, economy, and social structure to assemble operational plan achieving mission while strengthening desired political structures and earning local legitimacy
Evaluation	Critiquing higher or subordinate level's analysis in terms of both mission and grasp of local environment
	Judging the bias, interests, and utility of indigenous interlocutors' input about the AO

Table 9.5: Psycho-Motor Learning in Cultural Context

Psycho-Motor Domain of Learning	
Descriptor	**Characteristics**
Imitation	Repeating after instructor the rank-appropriate way to say and wave "hello"
	Practicing locally legitimate posture to demonstrate balance of Strength/approachability
Manipulation	Rehearsing staff-level introductions to indigenous partners for team/mission legitimacy
	Conducting mock informal informational interviews with role-players, repeating exercise based on feedback from role players
Precision	Without need of guidance from higher, performing cordon-and-knock on house with multi-generational family, separating males and females; according culturally proper treatment to male leaders, including phrases in local language; deploying female personnel appropriately
	As small unit, present hard target earning confidence of local families/deter escalation of force by local agitators
Articulation	Immediately assuming appropriate social intimacy with local interlocutors; demonstrating comfortable use of pleasantries, honorifics, spatial dynamics, expressions of emotion and solidarity
	Shifting among adversarial/conciliatory negotiation styles given significance of topic, public vs private nature of interaction, seniority of the local interlocutor and his evolving responses, considering overall impression one wishes to cultivate of strength and personality
Naturalization	Conducting proceedings for local communities in such fashion as to exhibit complete mastery of culture group's understandings of social hierarchies, beliefs, and unacknowledged biases, while exhibiting fluency with local ways

Table 9.6: Affective Learning in Cultural Context

Affective Domain of Learning	
Descriptor	**Characteristics**
Receiving	Exhibit particular focus on cultural topics related to individual Marine's tactical functions
	Belief that cultural training or cultural curricula in PME are legitimate aspects of learning
Responding	Active participation in seminars/scenario-based training, volunteering through questions/demonstration of skills
	Pursuit of individual reading and computer-based learning about the culture of the AO, and about cross-cultural interactions in general, even though it may not be directly related to the immediate information needed
Valuing	Believing that awareness of the culture of the AO is essential for tactical success and proper psychological and mental orientation to the mission
	Evaluating other Marines based on cultural conduct towards indigenous people and attitudes to the integration of culture in operational planning and execution
Organization	Preferring culture-heavy approach to planning and execution because it accords with what one feels to be their own natural approach
	Drawing parallel between culturally appropriate operational conduct, and military ethics, law of war, and human compassion befitting Marines
Value Complex	Un-self-conscious lifestyle viewing battlespace through Operational Culture Dimensions at tactical, operational, and strategic levels; is attitudinally disinclined to see it otherwise

Keying Operational Culture Learning to the Appropriate Domains

Up to now, we have discussed theories of learning as related to both the military as an organization, and to the nature of culture as a field of study and practice. With the exception of some concrete examples, our examination of these topics has not focused on specific geographic regions. As we indicated in the introduction to this chapter, it is important to distinguish between the study of specific regions and cultures, and the study of cultural concepts and principles, such as the Five Dimensions. These latter are applicable across regions and kinds of deployments. The realm of learning about broad cultural concepts, principles, and dimensions is professional military education (PME). By contrast, PTP requires specific regional focus.

Professional Military Education. In order for operational culture learning to occur successfully in PME, it must integrate multiple domains of learning. PME is not, however, concerned with learning about specific regions or particular culture groups. Rather, at every level, PME's primary focus is on developing rank-appropriate intercultural skills and conceptual grasp appropriate to any area of operations: It seeks to develop a frame of mind, or outlook, among Marines facilitating effective culture operations globally. As such, it is not concerned with "knowledge" in the cognitive domain, as informational learning and data recall do little to assist in skills and conceptual learning. Likewise, as so much of culture learning in the psycho-motor domain is by necessity region- and AO-specific, PME in the formal schools is an inappropriate environment to focus on it.

Instead, at the initial levels of structured military learning, be it enlisted Entry-Level Training and Marine Combat Training or Officer Candidates School and The Basic School, new Marines must encounter experiences, mentorship, and instruction which render them willing to learn about culture as a mainstream part of being a Marine and preparing for Marine activities. This requires learning in the affective domain up to the level of "receiving," which in turn will prepare them to "receive" and "respond" affectively to training in the PTP context when they are young enlisted or company grade Marines.

Cognitively, at the level of second lieutenants at The Basic School and sergeants in EPME, Marines need to gain a rudimentary grasp of the Five Operational Culture Dimensions. It is also necessary that they begin to understand the consequences of their actions at the ground level in terms of these dimensions—hence, cognitive "comprehension."

Figure 9.2: Bloom's Taxonomy in PME: ELT, TBS, Sgts

Domain	Ascending Levels of Learning					
	1	2	3	4	5	6
Cognitive	Knowledge	Comprehension	Application	Analysis	Synthesis	Evaluation
Affective	Receiving	Responding	Valuing	Organization	Values Complex	

As Marines proceed into the next levels of PME in Expeditionary Warfare School and as SNCOs, learning should elicit a desire to seek out more education about culture in military operations, both as a subject of study as well as from the standpoint of a Marine's duty as an educator and mentor to his younger charges. In this sense, curriculum developed appropriately, and teaching executed effectively, will cultivate "responding" among Marines. This affective "responding" will evolve gradually into "valuing," wherein a Marine will become committed to continuous culture learning as part of the responsibilities as a career Marine.

As a Marine will have by this point encountered non-regional specific culture learning in earlier PME experiences in conjunction with AO-specific culture training prior to deployments, s/he will now gain the cognitive ability to estimate future operational trends with respect to culture. This is a higher order of "comprehension," which will prepare him for "application" of the ideas and concepts animating this book to new contexts, situations, and questions, as appropriate to senior company grade and junior field grade officers, as well as staff sergeants and gunnery sergeants—meaning at the company and battalion staff levels.

Figure 9.3: Bloom's Taxonomy in PME: EWS and SNCOA

Domain	Ascending Levels of Learning					
	1	2	3	4	5	6
Cognitive	Knowledge	Comprehension	Application	Analysis	Synthesis	Evaluation
Affective	Receiving	Responding	Valuing	Organization	Values Complex	

Further cognitive learning is appropriate to the experience level and educational settings of mid- and top-level PME. At Command and Staff College, for example, majors should learn how to fully disassemble culture groups into the Five Operational Culture Dimensions, thus grasping the meaning and interrelationships among the environment; economic, social, and political structures; and beliefs and symbols—as relates to military operations in the planning and execution stages. Such learning foci correspond to "analysis" and "synthesis" in the cognitive realm, and relate clearly to the functions majors and lieutenant colonels will need to perform on battalion, regimental, and division staffs.

Successful learning at these stages, combined with successive PTP and experience gained through deployments, will permit particularly gifted commanders at the battalion level and above to judge, weigh, and balance the cultural, operational, and tactical issues at stake in his AO through the continuous Marine Corps Planning Process. As suggested above, successful integration of operational culture during the Mission Analysis, Commander's Battlespace Area Evaluation, and COA Comparison and Decision phases will permit the commander to approach the "evaluation" level of the cognitive domain, relying on specific subject matter expertise within his planning team.

Higher levels of "synthesis" and the realm of "evaluation" are, however, more suited to the experience and study of the gifted FAO who, when part of the Operational Planning Team as the subject matter expert, can apply what s/he has learned even beyond their primary region.

Figure 9.4: Bloom's Taxonomy in PME: Mid- to Top-Level

Domain	Ascending Levels of Learning					
	1	2	3	4	5	6
Cognitive					Synthesis	Evaluation
				Analysis		
			Application			
		Comprehension				
	Knowledge					
Affective						Values Complex
				Organization		
			Valuing			
		Responding				
	Receiving					

In the same vein, top-level school and further military educational opportunities can build upon experience to foster the basics of an internal value *system* integrating operational culture understanding with military considerations, when senior Marines encounter challenges at the tactical, operational, and strategic levels. Thus, PME at the uppermost levels can aspire to producing "organization" in the affective domain, principally because likely future deployments will require it, as the case studies in Chapter Eight have implied.

A final note about PME is in order here. We have emphasized the need for this kind of learning to be non-region focused—PME should develop Marines who are competent as planners and operators no matter where the AO. However, there is a need for some regional context to conceptual learning. Throughout this text, in fact, we have used examples from specific regions or culture group, in order to illustrate broad cultural dynamics. Likewise, applying principles of culture to specific regions in PME permits the learner to understand those principles better. In this case, regional learning is not primarily for the purpose of a profound, deployment-ready capability in any region itself, but for the sake of better understanding and applying the concepts and principles of Operational Culture in general.

Pre-Deployment Training (PTP). By contrast, PTP must be region focused by its very nature, and its goal is deployment-ready capability in the current operating environment (COE). For specific AOs, it is necessary as a basis to encounter basic cultural information, recall it, and enumerate it. This factual information is found in culture guides and materials associated with cultural intelligence, and corresponds to the "knowledge" component of the cognitive domain. As such, it is a quite basic initial foundation for deeper skills across all three domains.

Real capability in cultural PTP can only result, however, from higher-level learning. Cognitively, this means explaining cultural facts and processes in an AO in a fashion related to rank and billet, and then going on to consider consequences of Marine actions. For more advanced learners, or for those whose deployments will require more intense interaction with indigenous people, they will need to learn how to adapt their understandings of the culture to new likely scenarios through application of principles they have begun to derive.[11]

For example, a Marine might have learned about the social structure and its relationship to the civilian administration of a disaster-affected area, but he or she might have to adapt that learning to a mission focusing on assisting the local police or army in disaster relief and counter-narcotics. Thus, in the cognitive domain of PTP, Marines need to develop skills in the "knowledge," "comprehension," and "application" areas. To expect capabilities at the higher cognitive levels is not realistic, given time constraints for pre-deployment training and likely skills of in-unit and external subject-matter experts.

Because functioning with cultural competence also requires performing certain tasks, **psycho-motor learning is at the core of PTP**, particularly for the younger-to-mid-level Marine. This means learning how to replicate culturally appropriate actions that have been demonstrated through seminars and computer-based settings. As in other military training, this learning will often be through trial, error, and evaluation by qualified personnel—Marines will have to perform the task of a town hall meeting, or establishment of an observation post in an occupied residence, several times in training until they do it in accordance with performance standards.

Some of these tasks will need to become habitual, in that a Marine will need to learn how to do them several times in different settings with a proficiency convincing to indigenous people—for example, he or she will need to know how to say hello with proper combination of words, body language, etc., to elicit the desired response and attitude on the part of indigenous people. However, a Marine in most cases will not need to, or be able to, perform tasks so smoothly as to show no hesitation.

Therefore, with respect to the psycho-motor domain, while PTP should target "imitation" and "manipulation" in a mission-, rank-, and function-appropriate way, it is unrealistic to expect Marines to achieve "precision" or higher level skills in PTP. "Precision" in psycho-motor tasks can be aspired to by the foreign area officer (FAO) who combines academic learning about a particular region with extensive experience there.

For **pre-deployment** training's lessons to truly endure throughout a deployment, training **must also touch upon the affective domain.** Here, a Marine must be willing to actually pay attention to training, and "take it on board." Further, they need to be taught in a way that ensures active participation in training, so that they actively seek out such training. After successive exposures to PTP where material covered demonstratively assists in missions, Marines will begin to see intrinsic value in such learning. This means that PTP must embrace the affective domain, to ensure that Marines "receive," "respond" to, and "value" operational culture. Higher level affective learning is not necessary in the PTP environment.

Figure 9.5: Bloom's Taxonomy in PTP

Domain	Ascending Levels of Learning					
	1	2	3	4	5	6
Cognitive	Knowledge	Comprehension	Application	Analysis	Synthesis	Evaluation
Psycho-Motor	Imitation	Manipulation	Precision	Articulation	Naturalization	
Affective	Receiving	Responding	Valuing	Organization	Values Complex	

The value of PTP culture learning reaching the "application, "manipulation," and "valuing" levels of Bloom's taxonomy is not limited to training for tactical functions in a field exercise context at the battalion and lower level. For MAGTF staffs at the MEU through MEF level, accomplishments in the domains of learning express themselves best at mission readiness exercises (MRX).

Here, the cognitive ability to assess the operating environment in terms of the Five Operational Culture Dimensions can be fully developed to the point of "application," and tested as well. Likewise, the process of planning and conducting MAGTF staff activities creates the shared situational awareness that drives the affective learning process of staff-level PTP to reach "valuing."

Finally, simulated interaction with coalition and indigenous partners is an essential aspect of psycho-motor "manipulation." Therefore, MRX planners and participants must include culture learning through both seminar and simulation, lest the battle staff itself be out of sync with subordinate maneuver elements with regards to Operational Culture Learning.

To consider the matter from an alternative perspective, in the PTP realm, there are particular substantive cultural learning "building blocks" to integrate into unit as well as individual training. These building blocks overlap with the cognitive and psycho-motor domain. They also establish a progression of subject matter learning as a Marine or unit approaches deployment. Furthermore, these building blocks incorporate attention to the Five Operational Culture Dimensions, and how they apply in the specific AO. These "building blocks" of operational culture learning include:

Conceptual Grasp of the Principle: Understanding the reasons for, and implications of, a cultural concept, as well as the implications of our actions for that concept.

Understanding the Operational Culture Dimensions at Work: Detecting the components of specific Operational Culture Dimensions present in the AO, as well as their interactions, with respect to the cultural concept under study.

Concrete Information: Data about how a specific culture group within the AO demonstrates behaviors, structures relationships, and manifests beliefs regarding the cultural concept at hand.

Measurable Skills: Psycho-motor tasks the Marine must perform when operating among people from a specific culture group. These measurable skills provide individual training standards (ITS) and collective training standards (CTS), aligned with the culture, rank/billet of operator, security considerations, and overall mission goals.

Example: Culture Learning Building Blocks in PTP – Gender

When learning about gender in a Muslim Middle Eastern AO during pre-deployment training, **conceptual grasp of the principle** would mean understanding that relations between Muslim males and females, and particularly between "stranger" males and local females, are very different from those found in the U.S.. Further conceptual grasp would lead one to see that our treatment of indigenous women in what is considered a "too-aggressive" fashion will result in negative attitudes towards Marines as harsh, lascivious, and disrespecting of females, who collectively represent national honor and purity.

Understanding the Operational Culture Dimensions at Work would mean learning that in this Middle Eastern AO, the change in work patterns and economic conditions (Operational Culture Dimension Two: Economy) meant that more women entered the formal economy, thus undermining indigenous males' sense of provider-manhood, and increasing the incidence of male-female contact (the "gender" component of Operational Culture Dimension Three: Social Structures).

Over time, this economy-driven change in the social structure further increased males' desire to protect and seclude females related to them (the "norms, mores, and taboos" component of Operational Culture Dimension Five: Beliefs and Icons), while it also drove more females to dress in a more conservative fashion, thus permitting greater interaction with non-related males outside the home. Likewise, because previous imperialist powers focused on "liberating" women as a main component of their efforts to "modernize backward colonies," local (male) leaders will likely be quite sensitive to how U.S. forces act towards indigenous women, while those same indigenous women might resent being "liberated" from "restrictions" they themselves assumed.

Concrete information would mean knowing that in the more urbanized area of a particular country, the strictures regarding male-female interaction were lessened due to westernization, wealth, or higher incidence of non-majority religion populations. Concrete data would also tell the operator that the influence of patriarchic extended families and non-exposure to Western ways meant that the strictest gender-mixing rules applied.

Measurable skills would ensure that the Marine knew how to walk near, speak to/about, motion towards, and look at women in a particular place and time; was able to consider all the tactical functions that may be influenced by gender relations in that particular place; and knew whether and how to alter tactical functions in order to still accomplish mission according to commander's intent.

It is helpful to think about this all as a process applying through time, where the starting line is the beginning of a pre-deployment work-up, and the finish line is the early stages of the deployment itself.

Figure 9.6: Cultural PTP on Road Through Deployment

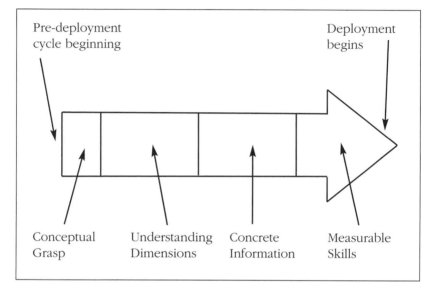

One might think that conceptual grasp, understanding societal conditions, concrete knowledge, and measurable skills are unique to one rank level or place in command hierarchy. This is not the case.

These building blocks of operational culture knowledge apply to all personnel. Given the challenges of the contemporary global operating environment, it is no longer enough for the Marine to know what to do/not do, yet with no concern for why they must/must not do it.

> *Example: Learning Building Blocks Apply to all Ranks*
>
> In the Central African Republic, a French Marine lance corporal on patrol far from base had been instructed not to search clergy. When he encountered a local missionary, not only did he refrain from searching him, but he initiated a rapport. This rapport yielded much intelligence about local networks of power and influence, and improved the unit's relationship with the indigenous people. If the LCpl had only been instructed not to search the clergy, but was not informed that the clergy have important family, economic, and political relationships, he might not have established the rapport that benefited the entire company.

As Chapter Eight implies, there are a few additional considerations that need to be addressed in the process of conceptualizing and executing cultural PTP. Though not related directly to the issue of learning domains, we raise them here, as they help to produce the **framing questions in developing cultural PTP.**

The identity of the Marine in terms of rank and billet. The identity of the Marine in functional and rank terms, and the identity of his/her unit in terms of location and role, are important considerations in PTP. Only after considering these matters can an instructor develop and execute PTP suited to the individual Marine and unit receiving training.

Most basically, in any operation, a corporal or sergeant will not be doing the same thing as will a lieutenant colonel or colonel. Thus, in terms of the Operational Culture Dimensions, the former will need different kinds of knowledge and skills than will the latter. Significantly, it is *not* a matter of *more or less*, but *different.*

> *Example: Rank, Billet, & Operational Culture Learning in PTP*
>
> As an example of different learning requirements at different levels and functions, in a counterinsurgency environment, a squad leader will need to know what kind of body posture and weapons positioning communicates strength, security, threat, lack of aggressive intent, respect, and openness to passers-by, fathers of families, etc.

Example: Rank, Billet, & Operational Culture Learning in PTP; continued.

He will also need to know gender-appropriate performance at vehicle check points and house searches, as well as what the culture's social structures mean for Marine interactions with family leaders, females, youth, and religious representatives.

Likewise, if working with local police or soldiers, the young Marine will need to grasp culturally-coded patterns of officer interactions with enlisted, and the importance of age, rank, and wealth in a small unit environment. They will also need to understand how environment and economy—levels of education, health, and nutrition—influence the tactical capability and attentiveness to detail of local security personnel. Marines in these roles will also need to be able to use a certain category of phrases in the local language, in addition to non-verbal symbolic communication appropriate to interaction at his level of society.

While the commander at the battalion level or above does not require "in the weeds" awareness of these matters, he does need to know the overarching environmental, economic, and social parameters which have caused the behavior and actions that his units will encounter. Not only will he need to grasp history, but he will require an even closer understanding of the local memory of it. This is because his likely interlocutors will be influenced more by the latter in a negotiation and planning context.

And, though he need not be fully versed on all the AO-specific manifestations of the Operational Culture Dimensions, the commander must have a comprehensive grasp of how these dimensions work and interact conceptually, as well as a basic knowledge of how they play out in the battlespace. In this way, he can effectively conduct his Commander's Battlespace Area Evaluation (CBAE), determine his areas of interest and influence, and articulate his Critical Information Requirements (CCIR). He will then be able to direct his subordinates to fully examine the AO in terms of the Operational Culture Dimensions, through intelligence preparation of the battlefield and course of action development.

Given his conceptual grasp of cultural dimensions and his baseline understanding of the AO itself, the commander can then select the appropriate COA in terms of both culture and mission. The commander also needs to know the significance of certain kinds of Marine conduct, and the way to use certain culturally-attuned measures for leverage.

As for demonstrable skills, both the squad leader and the commander need to understand approaches to time and spatial relations. The former will display his understanding in the context of street stops.

Example: Rank, Billet, & Operational Culture Learning in PTP; continued.

The latter will instead display competency relative to meetings and negotiations, through space concepts; understandings of time; as well as non-verbal communication and the biographical backgrounds of key leadership
points of contact. Leaders at both levels will not only need to be proficient
in these demonstrable skills, but they will need to learn how to mentor
their immediate subordinates, both in the performance of these skills, as
well as in the affective integration of culture into the overall Marine combat schema.

Billet thus matters. An infantry squad leader needs to understand
certain particular aspects about the culturally-coded observable
human conduct, in order to quickly assess threat level and atmos-
pherics, etc. An NCO in the same battalion's intelligence section
requires a broader grasp of the cultural components and human
patterns at play, but not at the same "in the weeds" level as required by the squad leader.

Likewise, the Marine civil affairs sergeant requires a grasp of the
culturally-coded dynamics of inter-personal relations, methods for
consensus building, hyper-local governance issues, and approaches
to economic exchange. So, not only rank, but billet also is at play
in determining the operationally relevant components of culture.

The nature of the operation along the kinetic spectrum. From port
call, humanitarian action, and disaster relief, up through ther-
monuclear confrontation, different aspects and a different degree
of cultural consideration in the planning and operating process
apply. Some feel that high-intensity combat need not take a region's culture into consideration: Goals are concrete, kinetic, and
suited to force-on-force conventional metrics.

Still, in this context, culture does play a role, forcing Marines to
consider the opposing military's cultural warfighting proclivities.
As a few examples, these culturally-coded preferences in warfighting include how the civilian and military social structures, political
structures, norms and mores influence military hierarchies. Further, the cultural background of hostile and friendly foreign military

personnel will influence civil-military relationships, approaches to operational initiative, information sharing, preference for tactics, and many other aspects about the military profession. These factors about a foreign military's cultural complexion will still influence the kinetic phase and need to be considered in planning and operating.[12]

Metabolizing Operational Culture

Whether it addresses culture as a concept or cultures in specific geographical contexts, much of the writing and teaching in the field of culture has been undertaken by academicians, and goes on in the college classroom. As such, there has rarely been concern for functional capabilities as a learning outcome. However, for operational culture learning to be true "training" in a pre-deployment environment, it **must provide Marines with psycho-motor capabilities.** Marines must learn skills that can be demonstrated as the performance of tasks, which can in turn be evaluated by subject matter experts according to reasonably objective standards that mesh both tactical and operational needs with cultural considerations.

Of course, performance of tasks need to be related to rank and function, but the essential point is that **pre-deployment culture training which does not teach military intercultural skills, but confines itself to informational briefs or motivational exhortations, is not learning.** Marines need to deploy with concrete skills in the realm of culture.

These skills need to emerge from cognitive "comprehension" and "application," and will have to be flexible as to circumstance, particular missions, and rank as well as billet. As we indicated in the introductory comments of this chapter, an over-weaning focus on performing the same tasks without the cognitive under-girding does not create culture operators at the tactical level; rather, it results in unthinking automatons. Likewise, appropriate task performance in the psycho-motor realm needs to be influenced by the affective hunch of the sufficiently trained Marine.

This renders standardized criteria for task performance evaluation extremely difficult to determine, and perhaps a futile effort—different situations reward different ways of performing tasks. Here, a training program runs the risk of evaluating performance according to standards that are more related to what the evaluator wants than they are to the actual operating environment and the people in it.

This is no marginal point. The tendency of military training is towards the hyper-concrete, observable, gradable, and replicable. This training bias produces good results in the tactile sense: if a Marine is trained to fire his weapon in five different positions; disassemble it down to the last bolt against a stop-watch while blindfolded; clean it to the point of immaculateness; reassemble it again as fast as s/he disassembled it, and then expertly fire the weapon—s/he will be able to do it with equal excellence in Norway, the Philippines, and everywhere between. S/he will do it the same regardless of whether s/he is a lance corporal or a colonel, logistician or infantryman.

Understanding culture, and performing in different cultures, however, cannot be conceived of in that fashion. Every AO is different culturally, and the cognitive as well as psycho-motor requirements for effective culture operators change from AO to AO, rank to rank, and MOS to MOS. In training terms, this means that "tasks" will remain the same only if we define them so generically as to require interpretation into much more specific tasks as per AO and mission and billet. More than that, "conditions" will constantly change, and in many cases, "standards" will as well.

Therefore, overzealously **locking culture learning into a systematized straitjacket will force learning targets into sterile abstractions of the lowest common denominator**, permit mediocre subject matter expertise on the part of instructors, and—worse still—divert Marines' focus away from the broader understanding at the conceptual level which this book has communicated throughout.

Put differently, excessive concentration on applying conventional military training concepts to Operational Culture Learning might serve the structural and psychological needs of the military training establishment; more to the point, however, it will do violence to actual understanding of cultures on the part of Marines, and their functional competence in them. As suggested above, the resulting training and education programs will end up being more about "us"—the military establishment—than it will be about "there" and "them"—the actual, continually evolving cultural environment to which Marines will deploy, and the people in it.[13]

If, for example, experienced Marines talk of the need to "metabolize" operational culture in the fleet,[14] and build capabilities associated with "emotional intelligence," how can one systematize that learning, or test it through replicable, performable tasks? And who, aside from the indigenous person with whom the Marine actually interacts while in an AO and functioning in a specific mission, is suited to judge the effectiveness of the Marines' Operational Culture learning?

In short, culture must be integrated into Marine operations, and into education, training, and planning for those operations. Yet, **Marines will fail culturally if this topic is treated mechanistically, just like all other military skills.** This is because Operational Culture mixes military with non-military concerns, intuition with certainty, and knowledge with skills.

Who Teaches?

This raises the matter of who should instruct culture in the PME and PTP realm. While this is the subject of a separate study, we can here note a few points. **First,** the person who develops curriculum for culture and teaches it either regionally or conceptually must possess a superior level of subject matter expertise. Experience operating in foreign environments and among foreign peoples is a necessary—but insufficient—condition, since *experience is not to be confused with expertise.*

Rather, in the context of region-specific cultural training and education, the instructor must have thoroughly studied the region according to the Five Operational Culture Dimensions, and must be able to apply them in both conceptual and concrete terms to the audience s/he is teaching, in order to cover learning throughout the domains encountered above.

In the context of non-region specific PME, specialization in a region is still helpful, for without a concrete regional context and expertise, the Operational Culture Dimensions remain sterile and without meaning, for the teacher or student. Of course, this regional knowledge must be based on a profound grasp of the concepts covered in this book. Further, it must be combined with a clear understanding of the overall purposes of each level of PME.

As we saw in Chapter Seven, it is also important that the instructor have at least a rudimentary grasp of the language of the AO, to be able to discuss the social significance of language, and Marine use of language.

Second, since culture learning is neither briefing nor preaching, but rather teaching, the trainer-educator must know how to teach. In addition to attendance of the Formal Schools Instructors Course, observation of military and non-military instruction in subjects related to operational culture is essential to forming the basic skills foundation to teach in the cognitive, affective, and psycho-motor realm. In short, the trainer-educator must be a pedagogue.

These two points might seem obvious, yet they cut against the grain of some training and education in the military. Since every Marine is a rifleman, and every SNCO and officer a mentor and teacher, it is regularly the case that Marines with some knowledge of a topic will be called upon to teach it, usually from a scripted program of instruction. This idea can be summarized as "if he's in that billet, or if he's done it, he must know it, and if he knows it, he must be able to teach it."

Whereas this works for conventional military skills, it is not appropriate to culture learning. To use the example of language again, just because a person has been in France, that doesn't mean they know French. And just because they can speak French, that doesn't mean they know how to teach it.

Third, beyond subject matter expertise and pedagogical capability, the instructor must be able to integrate culture into the military schema. He thus must speak in an idiom and use examples that are relevant to activities and practical concerns of Marines—this means grasping the Marine Corps (audience), understanding the reason Marines are deploying to an AO (mission), and diagnosing which aspects of the local environment are relevant to the particular functions of the Marines attending instruction (rank and billet). This, of course implies a final requirement: that the instructor be inclined to continuous learning, both about the realm of operational culture and the evolving Marine mission. Both kinds of learning drive the conscientious subject matter expert to listen to Marines.

Conclusion

In this chapter, we articulated a triangular relationship linking culture, military needs, and the science of learning. We also emphasized that culture learning for operational utility is a shared responsibility among schools, units, and individual Marines. While sharing this responsibility, Marines must ensure that training or education focus on the cognitive, psychomotor, and affective learning appropriate to one's rank and billet. Indeed, affective learning resulting in emotional intelligence may be the essential bridge between culture and military applications. Finally, while the military traditionally favors a method of developing curricula and assessing learning suited to concrete, quantifiable skills, the nature of Operational Culture requires a more nuanced method that accounts for the diversity of human phenomena and cultural behaviors, even in a single battlespace.

Notes

[1] Marine Corps Doctrinal Publication 5, *Planning* (Washington, DC: United States Government as Represented by the Secretary of the Navy, 1997), 15, 17.

[2] For discussion of needed approaches to learning for "post-modern" operational environments of the future, see LtGen Sir John Kiszely, "Post-Modern Challenges for Modern Warriors," published in *Shrivenham Papers* Number 5, December 2007: http://www.da.mod.uk/colleges/csrc/document-listings/monographs/shrivenham_paper_5.pdf. For broad coverage of the range of issues related to training, education, and learning for current and future operational envirnments, see "Conference Proceedings – Pedagogy for the Long War: Teaching Irregular Warfare (29 Oct – 1 Nov 2007)," compiled and edited by Barak A Salmoni (Marine Corps Training and Education Command). In particular, see Col Bill Monfries, "Initial Officer Training in the Australian Army," 84-85, and idem., "Excerpts from Key-Note Comments," 93-97; in addition to panel recommendations, 98-114.

[3] The literature on Bloom's Taxonomy has continued to evolve over the past fifty years, with additions and revisions. For a small sample, see Benjamin Bloom, *Taxonomy of Educational Objectives, Handbook 1: Cognitive Domain* (Boston, MA: Addison Wesley Publishing, 1956); Benjamin S. Bloom, David R. Krathwohl, and Bertram B. Masia, *Taxonomy of Educational Objectives: Book 2 Affective Domain* (New York: Longman, 1999); Bryant J. Cratty, *Psycho-Motor Behavior in Education and Sport* (Charles C. Thomas Pub Ltd, 1974); also see Lorin W. Anderson and David R. Krathwohl, *A Taxonomy for Learning, Teaching, and Assessing: A Revision of Bloom's Taxonomy of Educational Objectives* (Boston, MA: Allyn and Bacon, 2000); E.J. Simpson, *The Classification of Educational Objectives in the Psychomotor Domain* (Washington, DC: Gryphon House, 1972); Anita Harrow, *A Taxonomy of Psychomotor Domain: A Guide for Developing Behavioral Objectives* (New York: David McKay, 1972).

[4] William D. Wunderle, *Through the Lens of Cultural Awareness: A Primer for U.S. Armed Forces Deploying to Arab and Middle Eastern Countries* (Ft. Leavenworth: Combat Studies Institute Press, 2006), 10-12.

[5] See J.F. Kihlstrom and N. Cantor, *Social Intelligence* (London: Cambridge University Press, 2000).6 See H. Gardner, *Frames of Mind: The Theory of Multiple Intelligences* (New York: Basic Books, 1983); David Goleman, *Emotional Intelligence* (New York: Bantam Books, 1995); P. Salovey and D.J. Sluyter, eds., *Emotional Development and Emotional Intelligence: Educational Implications* (New York: Basic Books, 1997).

[7] See D.A. Kolb, *Experiential Learning: Experience as A Source of Learning and Development* (Engelwood Cliffs, NJ: Prentice-Hall, 1984); P.C. Earley, "Redefining Interactions Across Cultures and Organizations: Moving Forward with Cultural Intelligence," *Research in Organizational Behavior* 24, 2002, 271-99; D.C. Thomas and K. Inkson, *Cultural Intelligence: People Skills for Global Business* (San Francisco, CA: Berrett-Koehler, 2004). Also see D.C. Thomas, G. Stahl, E. Ravlin, E. Pekerti, et al., "Cultural Intelligence: Domain and Assessment," *International Journal of Cross-Cultural Management* (Forthcoming: 2008).

[8] Barak Salmoni, "Debrief of 5th CAG Falluja CMOC Director," Falluja, Iraq, August 2005; "Debrief of I MEF (fwd) G-2," interview by Dr. Barak Salmoni, al-Anbar Province, Iraq, October, 2006; *Final Report: I MEF Headquarters Group* (Quantico, VA: Marine Corps Center for Advanced Operational Culture Learning, 2006); *Final Report: Team Phoenix* (Quantico, VA: Marine Corps Center for Advanced Operational Culture Learning, 2006). For exploratory discussions of social and emotional intelligence in a military context, see LtCol Sharon M. Latour, USAF, and LtGen Bradley C. Hosmer, USAF (ret.), "Emotional Intelligence: Implications for all United States Air Force Leaders," *Air and Space Power Journal* Winter 2002: http://www.airpower.maxwell.af.mil/ airchronicles/apj/apj02/win02/latour.html; Holly Livingstone, Maria Nadjiwon-Foster, Sonya Smithers, "Emotional Intelligence and Military Leadership," Canadian Forces Leadership Institute, 11 March 2002: http://www.cda-acd.forces.gc.ca/CFLI/engraph/research/pdf/08.pdf; also see LtGen Sir John Kiszely, "Post-Modern Challenges for Modern Warriors."

[9] See Brian R. Selmeski, *Military Cross-Cultural Competence: Core Concepts and Individual Development* Royal Military College of Canada, Centre for Armed Forces & Society. Occasional Paper Series #1: 16 May 2007; Allison Abbe, "Developing Cross-Cultural Competence in Military Leaders," in *Pedagogy for the Long War: Teaching Irregular Warfare – Conference Proceedings*, 71-72; idem., with Lisa M.V. Gulick, and Jeffrey L. Herman, *Cross Cultural Competence in Army Leaders: A Conceptual and Empirical Foundation* U.S. Army Research Institute for the Behavioral and Social Sciences, Study Report 2008-01, October 2007.

[10] Dick Camp, *Leatherneck Legends: Conversations with the Marine Corps' Old Breed* (Osceola, WI: Zenith Press, 2006); Larry Smith, *The Few and the Proud: Marine Corps Drill Instructors in Their Own Words* (New York: W. W. Norton, 2007); Thomas Ricks, *Making the Corps* (Scribner, 1998).

[11] See Col Bill Monfries, "Initial Officer Training in the Australian Army," 84-85, and idem., "Excerpts from Key-Note Comments," 93-97.

[12] For more on culture learning in PTP, and the specific Marine Corps Approach executed by the Center for Advanced Operational Culture Learning, see Barak Salmoni, "Advances in Pre-Deployment Culture Training: the USMC Approach," *Military Review* (Nov-Dec 2006). For advances in the U.S. Army, see, in the same issue, Maj Remi Hajjar, USA, "The Army's New TRADOC Culture Center."

[13] See Papers by Allison Abbe, Charles Kirke, Col Bill Monfries, and Paula Holmes-Eber in "Conference Proceedings – Pedagogy for the Long War: Teaching Irregular Warfare (29 Oct – 1 Nov 2007)," compiled and edited by Barak A Salmoni, as well as related panel recommendations.

[14] Personal communication, Col B.P. McCoy, April 2006; "Debrief of I MEF (fwd) G-2;" *Final Report: I MEF Headquarters Group.*

Conclusion

This book has shown that Operational Culture considerations are relevant across the geographic and operational spectrum. Particularly in the hybrid and irregular operations characterizing the contemporary and future expeditionary environment, cultural considerations will be paramount. As such, *Operational Culture for the Warfighter* has pursued two major objectives. Our first objective has been to provide clarity of thinking with respect to defining and understanding culture—as a warfighting and operational competency relevant at all ranks and levels of command. This was broadly the focus of Part I.

Our second objective, and the focus of Part II, has been to provide a systematic, intellectually rigorous framework for understanding cultures globally, in a way attuned to operational needs. We did this by presenting and examining the Five Operational Culture Dimensions, including with them questions that should be asked by Marines in an evolving fashion through the phases of planning and execution. In Part III we demonstrated that these Dimensions indeed do animate the drivers of conflict in regions of strategic interest to the U.S., and to which Marines might deploy. We then suggested a way to think about learning in the realm of culture, to ensure that training and education—as an individual Marine, in schools, or in the fleet—provide the needed skills and mental orientation to benefit operations.

In preparing this text, the authors were struck by **two recurrent phenomena. First:** military organizations—and particularly the U.S. Marine Corps—are themselves culture groups, which can be analyzed across most aspects of the Five Operational Culture Dimensions. That is why we opened this text with METT-TC-MC, with C-MC standing for **C**ivilian culture *and* **M**ilitary **C**ulture considerations. **Second**: key concerns regarding culture in the military domain are mirrored by Marine Corps doctrine's concerns, particularly with respect to planning and warfighting. In this conclusion, therefore, we will begin by discussing militaries as cultures, taking advantage of knowledge the reader will likely already possess. We will then complete our inquiry into *Operational Culture for the Warfighter* through a review of some central conceptual considerations in the doctrinal literature on planning and operating, as a vehicle to summarize central ideas in making culture of utility to the warfighter.

Militaries as Cultures: The Marine Corps

Scholars of armed forces have come to think of "military culture… as the 'bedrock of military effectiveness' because it influences everything an armed service does."[1] Likewise, researchers with military backgrounds have come to see service cultures as the main factor in facilitating or inhibiting change—educationally, structurally, or operationally.[2] Numerous works, both popular and academic, have emerged of late to examine military culture.[3]

An American soldier, sailor, airman, or Marine will likely consider it obvious that militaries have cultures. "Army values," "Naval traditions," "Marine leadership principles and traits," etc., are all explicitly cultural markers of our services. Indeed, we frequently encounter Marines who explain why they do what they do by saying "it's a Marine culture thing." For this reason, this book has at times illustrated cultural phenomena by referring to the Marine Corps itself.

Still, in understanding their own service, or in coming to grips with the ways of other U.S. services and coalition partners, Marines require a guide to systematically analyzing militaries as culture groups. This is possible to do, through the very same Operational Culture Dimensions applicable to the battlespace as a whole. "Militaries as cultures" could easily be the subject of an entire book. Here, therefore, rather than a systematic investigation, we will very briefly survey the topic, using the Marine Corps for examples.

Since the authorization to establish two battalions of Marines in 1775, the Corps has functioned in two distinct physical environments: a maritime and amphibious existence coupled with austere and grueling ground conditions. Originally intended to be used as boarding parties, sharpshooters, and ship-board police, the Marine Corps later became a port and sea installation guard force. During WWI, it began to engage in large scale conventional military operations akin to those of the U.S. Army. Still, even after that, the Corps retained maritime and amphibious operations as an environmental constant. From Tarawa to Iwo Jima in WWII, the Marine Corps was a sea-based assault force, both doctrinally and in practice.

This maritime condition has continued, from Inchon (1950) to the current generation—in fact, the Marine Expeditionary Unit (MEU) is the most recent doctrinal and operational incarnation of a maritime environmental reality. This water-based aspect of the Corps' environment is clearly translated into the language and identity of the Corps. For example, Marines refer to land based facilities using maritime terms. From "head" to "deck" to "overhead," "ladder" to "bulkhead," Marines often cultivate an identity of being ship-borne, even while on land.

The Marine Corps is also an expeditionary land force, ready to operate in any land environment, from desert to mountain, from the Arctic to the tropics. Marines expect to operate under extremely difficult physical conditions, with few amenities. As a result, the Marine Corps has developed a culture of hardiness, extreme physicality, and austerity. In Boot Camp and Officer Candidates School, Marines are selected and trained for physical strength, endurance, resilience, and tolerance for pain and discomfort, going through one of the most rigorous training programs of any military in the world. They quickly learn to operate in Spartan environments using minimal equipment, and to be prepared to adapt to whatever situation they may face.

Even more than other U.S. services, Marines engage in, and are socialized to value, hard labor—as one SNCO commented, "working stupid-hard beyond the necessary."[4] Physical exercise—perennial "PT"—is quite prominent in the Marine's day, both as personal development, and as articulation of values. Sweat, and its smell, are thus quite familiar to the Marine and those who work with them.[5]

This cultural notion of austerity and hardiness combines with the Marine maritime legacy through the tightness of space associated with vessels. Quonset huts and squad bays are small, so Marines must "use, reuse, and reuse again," while being perennially tidy and "squared away"—able to "pack our trash," as scarcity of resources and space, as well as the expeditionary imperative, permit nothing else. In a rear environment, this condition translates into norms such as "hot-racking" computers, whereby several Marines use the same computer with separate log-ins—and this is perceived as normal, even when there are additional computers waiting in boxes.

Recently, the dominant aspect of a Marine's physical environment has been the desert, from Afghanistan to Iraq. But this reinforces the same practices and norms of austerity. For example, Marines live in "camps," not the Army's "forts," and operate from "combat outposts," as opposed to "forward operating bases."

The Marine Corps' cultural identity as an austere service is also reflected and reinforced in its economic relationships with the larger U.S. military structure. As a sub-component of the Naval Service, and in a competition for resources with both the Navy and the much larger U.S. Army, the Marine Corps has regularly faced shortages of basic items, and has often had to content itself with hand-me-downs from other services. This was the case before WWII, but events then seared a sense of scarcity onto Marine historical memory. During the second day of the Guadalcanal campaign, Navy vessels departed the area—along with most of the Marines' logistical and heavy equipment support. For the next four months, Marine forces had to use captured Japanese equipment and food to survive.[6] This historical memory fed into the ways Marines attitudinally encountered recurrent shortages in the 1950s and 1970s.

This scarcity and austerity continue to be seen today in conscious as well as unconscious practices. For example, from file folders to ammunition magazines, material items are used over and over, often to the point of destruction, before replacement. And, in many places, Marines go without—most comically, in restrooms (heads), where there are frequently no paper towels!

Their small size and sometimes unclear position within the U.S. military's social and political structure has further reinforced the notion of scarcity. Throughout its history the Corps has repeatedly had to justify its very existence. From the 1920s through to the 1960s, many in the executive branch of government, Army, and Air Force questioned the need to have a Marine Corps as a separate service. The Corps' successes in WWII, particularly at Iwo Jima and Okinawa, ironically hurt its image as a separate, unique service: It could take on high intensity ground combat—as could the Army—and it possessed carrier- and ground-based aircraft—as did the Navy and the new U.S. Air Force.[7]

This ambiguous economic, social, and political position within the U.S. military has led to a belief that the Corps as institution is always under threat, and must show that it can always do all things, better than other services. As a result, Marines have been characterized as exhibiting a clannishness and healthy skepticism towards other services—and even at times for the civilians they protect.[8] The cultural attitude that Marines are separate, standing above all others is reflected in their icons. The globally most recognizable Marine icon is the flag raising atop Mt. Suribachi on Iwo Jima, on 23 February 1945. Marines—and sailors—who landed by naval vessels raised an American flag, both culminating and expressing extreme physicality, literally raising the flag beyond anyone's individual reach.

As the afore-going shows, the Marine Corps environment and its role within the U.S. military structure have implications for its beliefs, icons, and historical memory. This environment—or the way it is re-remembered daily—also has implications for the Marine Corps' own internal economy—or ways of structuring labor and exchange—and social as well as political structure. Most basically, one could argue that the Marine Corps exhibits a mixture of egalitarian distribution, informal exchange, and industrial production. Likewise, the Marine Corps as an institution, and individual Marines, often exhibit industrial-era attitudes to human resources and manpower uses. Finally, though more egalitarian than other U.S. military services, the imperatives of a mass-manpower, military-bureaucratic organization create social structures and political relationships that are quite hierarchical when compared to civilian society, but which are characterized by a fair amount of formal and informal mobility.

As for distribution and exchange, given conditions of scarcity, in broad terms, no particular group of Marines is due more material goods than any other group by dint of status or position. Food is shared equally; residential quarters—while different from enlisted to company grade and field grade officers—are remarkably alike; and quality of clothing, medical treatment, and amenities of life is broadly the same throughout the Corps. To the extent that access to and quality of material differs, it is due to billet and function— the weapons, optics, and gear harnesses of Logistics Marines will differ from those of Force Reconnaissance Marines because of mission requirements, and not status.

At the same time as relatively egalitarian distribution, however, scarcity has elicited less formal practices. One of these is networks of informal acquisition and exchange. Senior SNCOs and Warrant Officers—or even LtCols—can create "mafias" to "arrange" acquisition of needed goods—particularly for matters directly related to warfighting or troop welfare for youngest Marines. It is here that the social structure feature of age grades, or cohorts of graduates form Marine Corps schools, has implications for economy. These informal networks are even used in personnel assignments, as people who know each other work through or outside the system to adjust its impersonal bureaucratic functioning. Over time, the ties of informal reciprocal obligation work to ensure that compensation is found.

A second, less formal practice is related to acquisition. Along with the explicitly socialized idea of "take only what you need," Marines have been known to hoard goods such as folders, toilet paper, MREs, magazines, etc., as the rainy day—or existential threat to the unit or Corps—might be around the corner. In its most extreme manifestation, this takes the form of going into a kit, or full set of gear, and yanking out those parts which one needs, leaving the rest behind.[9] As the Marine aphorism goes, "there's only one thief in the Marine Corps. Everyone else is just getting their stuff back."[10]

This collectivist, egalitarian distribution is also reflected in labor practices. Common work for a common cause is required for success in a world of scarcity, austerity, and danger—hence the explicit socialization to group-orientation: "there is no 'I' in team." Generally, all hands engage in field days, and most Marines, officers and enlisted, have to pull twenty-four hour guard duty on a randomized schedule.

This collectivism is also inspired by the de-personalization of an industrial organization. On the one hand, this is exemplified by the maxim, "every Marine's a rifleman." This attests to reality of manpower scarcity and need for adaptability, as well as the ethos that the Corps is a warfghting organism, no matter what the specialty of its individuals.

On the other hand, de-personalization reflects a late nineteenth- and twentieth-century bureaucratic, mass-based approach to manpower use: people are, like parts of a machine, interchangeable and moldable. This is not a criticism; rather, it reflects the institutional egalitarianism, denigration of individualism, and industrial machine-like nature of the Marine Corps—which, carved in the crucible from Belleau Wood through to Khe Sanh, differs from many twenty-first-century approaches to labor and individual worth.

Finally, the modern Marine Corps has become a mechanized, industrial organization. Tracked and wheeled movement; computerized land navigation; technology-based IED-defeating devices; and an increasingly sophisticated air wing have required higher levels of education and cognitive/psycho-motor problem solving skills—but have also been reflected in economies and ethos: Marines often buy over-large sports-utility vehicles, motorcycles, and gadgets, and are rarely found in hybrid automobiles.

Other aspects of the Marine Corps' social and political structure contradict egalitarianism, here reflecting its heritage as a military organization of the western European, British tradition, and as a social structure containing people of vastly different educational backgrounds and competencies. In this sense, the Marine Corps is highly stratified. Most obviously, there are officers and enlisted. Some of the former actually articulate it thus: "Enlisted do; officers tell [enlisted what] to do."[11] Though this might be an exaggeration, there is a clear difference between officers and enlisted Marines.

Other elements of stratification include military occupational specialty (MOS). The Marine Corps is a self-consciously combat arms—and infantry-centric—organization. For many, there is infantry, and then everything else. While the U.S. Army is much better known for MOS-based identity, in the Marine Corps too, there is a major differentiation between combat arms—highest in social status—and all else, beginning with intelligence, then logistics, then administration, etc. Within the combat arms, infantrymen consider themselves superior to armor and artillery—and a world apart from the aviator community. As an artifact of these differences, a lexicon of colloquial terms has emerged, both to identify oneself and others.[12]

Like other culture groups, the U.S. military reflects dynamics in larger American society—demonstrating again that culture groups are not impermeable compartments. Thus, the Marine Corps has grappled with matters of gender and ethnic minorities. For some time, both groups were consigned to lower social status and access to power. For female Marines, there are structural limitations imposed by U.S. law: excluded from the ground combat arms, they cannot access key billets of power in the Marine Corps. However, as they have moved into the Air Wing and intelligence, this has begun to alter. Furthermore, in recent operations in gender-differentiated areas such as Iraq and Afghanistan, female Marines have begun to demonstrate a unique utility, while also engaging in more infantry-like activities.

At both the tactical and operational level, therefore, female Marines have gained greater prestige, status, and influence—hence, a concrete reality has begun to change a cultural belief, reflected institutionally in the emergence of female General Officers up to the three-star rank.[13]

The same is true for ethnic minorities. Historically, the Marine Corps was largely white in its demographic complexion. Through to the 1960s, it was not a culture group congenial to African Americans or other ethnic groups.[14] Yet, as American society has changed, and as educational levels have risen among ethnic minority groups, more Marines from diverse ethnic backgrounds have demonstrated their indispensable skills to the Corps from Somalia to Afghanistan, Iraq to Indonesia.

Therefore, the Marine Corps has come to positively value diversity, both for its own sake, and because "every Marine's a rifleman" and an ethos of "mission, mission, mission." Indeed, in its self-representation—epitomized by recruitment drives—the Marine Corps has begun to highlight its diversity, and target specific ethnic groups.[15] The social and political structure of the Marine Corps has therefore begun to evolve. Battalion commanders from South Asia and regimental commanders from Hispanic backgrounds have joined African American General Officers at the three-star rank, and a Jewish Assistant Commandant of the Marine Corps.[16]

Finally, the Marine Corps possesses a sacred symbolic geography. "Holy sites" include Tun Tavern, Belleau Wood—"Bois de la Brigade de Marine"—Iwo Jima and its memorial in Washington D.C., as well as the Marine Corps Barracks at 8th and I streets in Washington. The Marine Corps also puts much focus on institutions—one might call them temples or churches—of explicit socialization and indoctrination; a caste of socializers (or priests); and a corpus of doctrine (scripture), accompanied by semi-canonical literature meant to continue the socialization process.

The temples are Boot Camp, OCS, School of Infantry, and TBS (as well as other PME institutions); the priests are the Drill Instructors, OCS instructors, and TBS instructors—as well as the SNCO Corps, as a sort of lay clergy that ministers both to younger enlisted Marines as well as to company grade officers. Marine Corps Doctrine, from the 1934 *Tentative Manual for Landing Operations,*[17] *1940 Small Wars Manual,* up through the Marine Corps Doctrinal Publications of the late 1990s and early 2000s,[18] is a Marine's corpus of scripture, while supporting books are the folk texts of the

Marine Corps religion,[19] which Marines read along with texts from the great sages of the Corps.[20]

Ultimately, every rank insignia and every ceremony, to include graduation from Boot Camp or OCS, promotions, changes of command, and culminating with the Marine Corps Pageant and Birthday Ball every 10 November—all these are the rituals that teach Marines what it means to be a Marine, and articulate identity, aspirations, and even anxieties to insiders as well as outsiders, ensuring the reproduction of ideology and social structure. And, as with other culture groups' belief systems, there are particular rites of passage, to include The Crucible for enlisted recruits, and "Quigley" as well as SULE II for officer candidates.[21]

Our brief survey of the Marine Corps as a culture has admittedly been "wave-tops" and incomplete. There are also elements of it with which an outside observer or Marine might justifiably disagree. However, its essential point has been to demonstrate that not only are militaries themselves cultures, but that they can also be systematically analyzed as such, according to principles aligned with the Five Operational Culture Dimensions. When contemplating operations against foreign militaries—or when planning interactions with other service branches or friendly foreign militaries—such an analytical approach should be borne in mind by the culture operator.

Principles of Military Doctrine: Congruence with Operational Culture

The correspondence between the Marine Corps' approach to planning and operating and our approach to Operational Culture can be seen in the following two passages from Marine Corps doctrinal publications:

> War is not a single problem, but a complex system of interdependent problems, the solution to each of which affects the outcomes of all the others.[22]

In war, no episode can be viewed in isolation. Rather, each episode merges with those that precede and follow it—shaped by the former and shaping the conditions of the latter—creating a continuous, fluctuating flow of activity.[23]

Parts I and II of this book, and in particular the examination of the Operational Culture Dimensions and their related case studies, have strongly argued the same point with respect to culture and the role of culture in military operations. We could have easily paraphrased the first quote above to say:

> *Culture* is not a single *and isolated component of the AO*, but a complex system of interdependent *dimensions*, the *dynamics and impacts of* each of which affect the *dynamics* of all the others.

Likewise, Chapter Eight clearly demonstrated that the second quote too can be paraphrased:

> In *culture groups*, no *dimension* can be viewed in isolation. Rather, each *dimension and component of culture* merges with those that precede and follow it—shaped by the former and shaping the conditions of the latter—creating a continuous, fluctuating flow of activity *in the culture groups* Marines will encounter.

Doctrinal Considerations are Operational Culture Themes. Given this background, it is significant that many principles and animating tenets of Marine corps doctrine also hold true for how a Marine should approach integration of culture into their thinking, planning, and execution. We review a few doctrinal considerations here.

Humans as Central to War. Marine Corps doctrine on warfighting teaches us that "the human dimension is central in war. It is the human dimension which infuses war with its intangible moral factors."[24] This idea serves as the foundational requirement for integrating culture into military learning, planning, and execution. It is also the cornerstone of the concept of "war amongst the people," the dominant mode of operations today and in the future. As we have shown in this book, failure to consider the "human dimension" will prevent commanders from arriving at full answers to questions such as "which factors are critical to the enemy? Which can the enemy not do without? Which, if eliminated, will bend him most quickly to our will?"[25] These are questions related to "the center of gravity" in military operations, and in the operations of today and tomorrow, that center of gravity will be the "human dimension."[26]

Leadership Involvement. No matter what the level or rank, the commander cannot be aloof from his battlespace. As doctrine reminds us, "Planning is a fundamental responsibility of command. The commander… must drive the process," as an active participant setting the overall intent and parameters, and making decisions.[27]

This principle is equally important to effective integration of operational culture in planning and operations. The commander should possess a thorough grasp of the influence of cultural factors within his AO, and articulate a vision, through commander's intent, for how to accommodate or use the cultural realities of the people and opposing forces in the battlespace. Fundamentally, absent the commander's demonstrated and consistent concern for cultural factors during planning, a planning team and subordinate commanders will encounter difficulty integrating it among "traditional" warfighting considerations, and the unit might have a lackadaisical approach to culture in training and operating.

This means leadership must also be integrally involved in Operational Culture Learning—both in their own experience of it, and in their prioritization of it among their subordinates. This is because commanders possess the ultimate moral authority to inspire subordinates to embrace affective learning, along with more traditional modes of military learning.

The Integrated Battlespace. As has been demonstrated throughout this text, neither the dimensions of culture nor their components are isolated from each other. In doctrinal terms, "events in one part of the battlespace can have profound and often unintended effects on other areas and events, [so] therefore a commander must always view the battlespace as an indivisible entity,"[28] since "no episode can be viewed in isolation."[29] As with warfighting, so with culture. Parts II and III of this book showed that events concerning one Operational Culture Dimension, for example the political structure, will almost certainly have profound effects on another Dimension, such as the economy or social structure. Likewise, the dynamics of a culture group in one geographic area will influence cultural dynamics in another area as well. Therefore, the Single Battle Concept is as cardinal a principle for culture as it is for one's overall scheme of maneuver.

Integrated Preparation. Just as the Five Operational Culture Dimensions are so integrated as to require the Single Battle Concept, proper planning using these dimensions needs to be systematic and thorough, meaning it must address all relevant dimensions throughout the planned operation's stages. One must apply a systematic, coordinated, and thorough approach,... through a planning team composed of subject matter experts... to consider all relevant factors."[30]

Planning involving culture must also be coordinated. A force therefore must consider not only their own actions, but the combined effects of several other components—friendly, neutral, and hostile—in the AO. Finally, effective integration of culture into operations and planning requires inclusion of cultural subject matter experts on planning teams.

A Common Mental Model and Operational Schema. "The process of planning itself should provide a common understanding of the nature of the problem and so support communication and cooperation. In other words planning is a way of exploring the situation... [as] a basis for unity of effort."[31]

By establishing a "common understanding" for how to look at culture groups in any AO and in any kind of operational context, the Operational Culture Dimensions permit Marines to "explore" an emerging situation together through a "unity of effort." By working through operational culture considerations as a planning team, that team will develop the necessary shared situational awareness.[32] In other words, culture-integrated planning as a team, conducted in conjunction with operational culture PTP, will develop the essential psycho-mental cohesion of the planning and operating staff. Through the vehicle of considering cultural considerations as an integrated team, that planning team will gain in self-awareness, battlespace awareness, and mission awareness, all of which will benefit the planning staff as it transitions into execution.

Shapes the Thinking While Embracing Fluidity. Part I of *Culture for the Warfighter* demonstrated the need for a disciplined way of thinking about culture and defining its core terms; we also showed how without such discipline, culture can have no utility to the warfighter. Marine Corps doctrine speaks in similar terms about planning. "Planning can provide a disciplined framework for approaching problems.... The key is to adopt a method that provides helpful structure without restricting judgment and creativity."[33]

Therefore, the systematic intellectual approach put forth in Part II can structure Marines' thinking about the cultural dynamics of any AO, and allow them to relate culture to the mission itself. However, as we have suggested throughout this text, and in particular in Chapter Nine dealing with training and education, understanding cultural dimensions in the battlespace is never meant to be mechanistic (if-this-then-that), nor is it meant to restrict the military judgment or intellectual creativity of the planner or operator. Rather, Operational Culture is meant to facilitate creativity through the proper framework. This well-framed creativity is essential for the kind of operations of today and tomorrow, which "take place in an atmosphere of uncertainty... [which] pervades battle."[34]

Warfare as Art. Though quantifiable knowledge and measurable skills are important in warfighting, "an even greater part of the conduct of war falls under the realm of art, which is the employment of creative or intuitive skills." It is the combination of the two kinds of intelligences that make Marines effective at "the situational application of scientific knowledge through judgment and experience."[35] The same could be said of culture in military operations. It is for this reason that this book, and the Five Operational Culture

Dimensions, have focused on concepts and emphasized the need for emotional intelligence in affective learning. Only this can bring art and an "intuitive ability" to the conduct of "culture operations" as we encountered them in Chapter Two.

Processes as Opposed to Procedure. "We should think of planning as a learning process—as mental preparation which improves our understanding of a situation."[36] "The true value... is the process itself, vice the product ('the plan') it generates.... [The process is] developmental stages, rather than steps, that are progressive in nature yet interwoven, overlapping, and often falling back on one another."[37]

This is very much the thrust of considering culture in planning, training, and operations. The true utility of Operational Culture and Operational Culture Learning is not in "products" such as "human terrain maps," manuals, or "cultural" tactics, techniques, and procedures. In fact, Operational Culture is not something that *is*; rather, it should be considered something one *does*. It is the process of thinking about the dimensions, their components, and the interwoven, overlapping nature of them all that provides an accurate—though one need not presume precise—vision of the relationship between proposed military actions and the cultural environment in which they will occur. The proper mental preparation, which emerges from combining operational culture in PTP and planning, will contribute most effectively to the overall developmental stages of evolving planning and operating.

Doctrine Misused is Operational Culture Misconstrued. In a similar fashion, the misuses of doctrine need to be borne in mind when applying Operational Culture. These misuses include:

Attempting to Forecast and Dictate Events. Though there may be a "natural desire to believe we can control the future," human beings—and the enemy—all have a will of their own.[38] Understanding the culture of an area does not permit a Marine to predict what indigenous people will do in anticipation of or response to Marine actions. In the same way, Operational Culture offers no explanatory silver bullet for why individuals or groups of people do what they do: A Marine's (and even an indigenous person's) cultural understanding of the AO will always remain partial, and neither Operational Culture Learning nor "cultural intelligence" provide the kind of blueprint to how other cultures think or act which could permit a Marine to predict or dictate events.

Imposing Clarity on an Unpredictable Environment. The "natural desire to leave as little as possible to chance" should not lull the Marine into forgetting that the battlespace and the people who make up the culture of it are all about chance.[39] One cannot plan out in too much detail matters related to the environment, economic system, political system, etc., because it is individuals' and groups' *interaction with* these cultural dimensions that result in conditions Marines face.

It is legitimate and proper to apply the Operational Culture Dimensions to an AO for the purpose of "anticipatory adaptation" through "evaluating potential decisions and actions in advance,... and thinking through the consequences of certain potential actions in order to estimate whether they will bring us closer to the desired future."[40] However, to think that culture can be "nailed down" in detail, or that one can fully map it—hence the problem with "human terrain" and "human terrain mapping"—is to misunderstand the nature of both culture and operations.

Scripting Processes—Friendly and Enemy. If operational planning or Operational Culture are considered as tools to script military action or peoples' activities—civilian or in uniform—Marine initiative will be lost, and we will fail to recognize indigenous peoples'—friendly neutral, or hostile—right of initiative or independent action. As such we will have disrespected our indigenous enemies and potential friends, to our operational peril.[41]

Inflexible Thinking with Emphasis on Procedures. While both Marine Corps Doctrine and the Five Operational Culture Dimensions seek to provide a disciplined framework to think about and engage in planning, educating, and operating, they are not intended as a rule book prescribing an order of steps. To remove flexibility of thought and action from planning or from the mental process of addressing the AO through the prism of Operational Culture is to "reduce those aspects of planning that require intuition and creativity to simple processes and procedures." Even appropriate procedures "naturally tend to become rigid over time."[42] Be it in relation to culture or warfighting, such an approach neglects the principle of "uncertainty and fluidity" enshrined in Marine Corps doctrine.[43]

We should consider these statements in light of concerns encountered earlier with respect to training and education. As we saw in Chapter Nine, there is a need for Operational Culture Learning to communicate specific concrete capabilities in the cognitive and psycho-motor domain. However, an excessive application of set procedures and rigid informational targets in the cognitive domain, as well as an insistence on scripted performance of tasks in the psycho-motor domain, will fail to grasp the nuanced role of culture and cultural competency in military operations. For this reason, we discouraged a mania for tasks, conditions, standards, learning objectives, etc., be it in PTP and PME, and encouraged instead a more flexible approach that would deter rigidity and lockstep thinking or acting.

***

Be it in high-intensity warfare or the kinds of operations requiring Marines to interact with foreign cultures in a variety of roles, "the problem will evolve even as we are trying to solve it."[44] Likewise, matters involving culture and military operations epitomize the "wicked" problem set, wherein problems are interdependent, and solutions to one part aggravate dilemmas in another part of the problem—at the same time as grappling with such problems is unavoidable.[45] Therefore, just as operational planning and decision-making are iterative, evolving processes continuing even during the course of operations, a well-framed but flexible consideration of culture must also continue throughout the phases of mission preparation, training, planning, and execution. That has been the goal of this book: a truly "operational" vision of culture.

Notes

[1] See Allan English, "The Influence of Our Military Culture on What is Taught in Our Professional Military Institutions," in *Pedagogy for the Long War: Teaching Irregular Warfare – Conference Proceedings*, edited by Barak Salmoni (Marine Corps Training and Education Command, 2008), 38.

[2] See Charles Kirke, "Change and Military Culture," in *Pedagogy for the Long War: Teaching Irregular Warfare – Conference Proceedings*, 36-37; R. M. Cassidy, "The British Army and Counterinsurgency: The Salience of Military Culture," *Military Review* (May 2005). At its base, John Nagl's *Learning to Eat Soup with a Knife: Counterinsurgency Lessons from Malaya to Vietnam* (Chicago: University of Chicago Press, 2004), is a study of how the historical memory of military operations, as well as organizational culture, influenced the nature of U.S. and U.K. operations, and their willingness as well as methods to learn in new circumstances. For more on U.S. military culture, see Adrian R. Lewis, *The American Culture of War: The History of U.S. Military Force from World War II to Operation Iraqi Freedom* (London: Routledge, 2006).

[3] For the Marine Corps, see Terry Terriff, "Of Romans and Dragons: Preparing the U.S. Marine Corps for Future Warfare," *Contemporary Security Policy* 28:1 (2007); idem., "Warriors and Innovators: Military Change and Organizational Culture in the U.S. Marine Corps," *Defence Studies* 6:2 (2006); Paula Holmes-Eber, "Mapping Cultures and Cultural Maps: Representing and Teaching Culture in the Marine Corps," in *Pedagogy for the Long War: Teaching Irregular Warfare – Conference Proceedings*, 73; idem., with Barak Salmoni, "Marine Officers' Imagined Self-Identities: Ethnographic Insights for Cultural/Psychological Training," 1st International Operational Psychology Conference, IDF Military Psychology Center, Herzliya, Israel, 01 Dec 2006. For an intimate and frank exploration of the culture of a Marine ground unit during high-intensity combat, see Maj Seth W.B. Folsom, USMC, *The Highway War: A Marine Company Commander in Iraq* (Dulles, VA: Potomac Books, 2006). A more journalistic work is Thomas Ricks, *Making of the Corps*. Journalistic reportage is often a strong source of learning about foreign military cultures. As examples, see Anna Politkovskaya, *Putin's Russia: Life in a Failing Democracy* (Holt, 2003); idem., *A Small Corner of Hell: Dispatches from Chechnya* (Chicago: University of Chicago Press, 2003). Militaries of the Middle East are covered with quite varying quality. Among the best are Eyal Ben Ari, *Mastering Soldiers: Conflict, Emotions, and the Enemy in an Israeli Military Unit* (London: Berghan Books, 1995); Uri Ben Eliezer, *The Making of Israeli Militarism* (Bloomington, IN: Indiana University Press, 1998), on the relationship among Zionism, the Israeli state, civil society, and the military; Mehmet Ali Birand, *Shirts of Steel: An Anatomy of the Turkish Armed Forces* (London: IB Taurus, 1991) covers the Turkish officer corps as well as his sources, access, and self-censorship would permit in the mid 1980s. Norville D'Atkine's "Why Arabs Lose Wars," *Middle East Review of International Affairs* 4:1 (2000) is of basic utility in descriptive terms, though with very little analytical merit, given its reliance on a Patai/"national character"

model. It has, in any event, been superseded by experiences and observations of U.S. and coalition personnel operating in Iraq. John Poole, *Tactics of the Crescent Moon: Militant Muslim Combat Methods* (Posterity Press, 2004), is of no utility, based on stereotypes, bias, and no empirical observation or research.
[4] Conversation with Marine Gunnery Sergeant, Quantico, VA, 2006.
[5] Conversation with battalion operations officer, al-Taqaddum Iraq, August 2005. Also see Evan Wright, *Generation Kill: Devil Dogs, Iceman, Captain America and the New Face of American War* (New York: Penguin, 2005).
[6] See Richard Frank, *Guadalcanal: The Definitive Account of the Landmark Battle* (New York: Random House, 1990); Samuel B. Griffith, *The Battle for Guadalcanal* (Champaign, Illinois: University of Illinois Press, 1963).
[7] See James A. Warren, *American Spartans – The U.S. Marines: a Combat History from Iwo Jima to Iraq* (New York: The Free Press, 1995), 97-118, 183-206.
[8] See Thomas Ricks, *Making the Corps* (New York: Scribners, 1998), 21-23, 162-63, 200-202, 274-79.
[9] This practice is known colloquially as "rat-fucking."
[10] Interview with Marine Aviator, al-Asad, Iraq, 2008.
[11] Discussion with Marine Officer, of prior enlisted and service academy background, Monterey, CA, 2005.
[12] As examples: "knuckle-draggers" is used for Marines generally, but particularly for infantrymen; "lanyard-pullers" can be used semi-affectionately for artillerymen; "go swing with the wing" can be used by ground combat arms Marines derisively about the aviator community, while aviators themselves can use it to signal their self-admiration. "Pogue" is a derisive term used to refer to administration Marines, though combat arms Marines can use it to refer to other combat arms Marines who have not left a rear-echelon environment for some time. "Leaf-eaters" is a new term used for civil affairs and psyops, with both negative and positive connotations. These are just a few examples.
[13] See James E. Wise and Scott Baron, *Women at War: Iraq, Afghanistan and Other Conflicts* (Annapolis, MD: U.S. Naval Institute Press, 2006); Kirsten Holmstedt, *Band of Sisters: Women at War in Iraq* (Stackpole Books, 2007). The covers of both these books depict female Marines in Iraq, in full combat gear.
[14] See Melton A. Mclaurin, *The Marines of Montord Point: America's First Black Marines* (Chapel Hill, NC: University of North Carolina Press, 2007). One could extend this comment to non-Christians: although Joseph W. Bendersky, *The "Jewish Threat": Anti-Semitic Politics of the U.S. Army* (Basic Books, 2001) is not about the Marine Corps, it does record certain comments and attitudes of Marine General Officers during the 1940-50s. See 411-18.
[15] See Richard Whittle, "Uncle Sam Wants U.S. Muslims to Serve," *Christian Science Monitor,* 27 Dec 06; on Marine Corps Recruitment Command's efforts, under then-BGen W.E. Gaskin, to target Jews of Middle Eastern extraction, see Shelomo Alfassa, "Is Uncle Sam Confused," *The Jewish Press,* 24 Oct 05.
[16] See Suzanne Kurtz, "Semper Chai: General Robert Magnus," *Hillel News* 21 Feb 2007: http://www.hillel.org/about/news/2007/feb/magnus_2007Feb21.htm.
[17] Re-titled *Tentative Landing Operations Manual* (Navy Department, 1935).
[18] These are the Marine Corps Warfighting Publications (MCDP) 1 to 5, and the associated Marine Corps Warfighting Publications (MCWP), as well as Marine

Corps Reference Publications (MCRP).

[19] A very short list of examples includes Kenneth W. Estes, *The Marine Officer's Guide*, 6th edition (Annapolis, MD: U.S. Naval Institute Press, 1996); *U.S. Marine Guidebook of Essential Subjects* (Dept. of Defense Legacy Resource Management Program / Apple Pie Pub, 1998); K.W. Estes and R.D. Heinl, *Handbook for Marine NCOs*, fourth edition (Annapolis, MD: Naval Institute Press, 1995); *Guidebook for Marines*, 18th rev. ed. (Quantico, VA: Marine Corps Association, 2001).

[20] A key example is LtGen Victor Krulak, *First to Fight: An Inside View of the U.S. Marine Corps* (Annapolis, MD: U.S. Naval Institute Press, 1999).

[21] See James B. Woulfe, *Into the Crucible: Making Marines for the 21st Century* (Presidio Press, 1999).

[22] *Marine Corps Doctrinal Publication 5: Planning* (United States Government as Represented by the Secretary of the Navy, July 21, 1997), 22. Henceforth MCDP 5.

[23] *Marine Corps Doctrinal Publication 1: Warfighting* (United States Government as Represented by the Secretary of the Navy, July 21, 1997), 9. Henceforth MCDP 1.

[24] MCDP 1, 13.

[25] MCDP 1, 46.

[26] For more on this in the context of the contemporary and future operating environment, see United States Joint Forces Command, *Joint Operating Environment: Trends and Challenges for the Future Joint Force through 2030* (Norfolk, VA: USJFCOM, December 2007); also see LtGen Sir John Kiszely, "Learning About Counter-Insurgency," *RUSI Journal* 152:1 (December 2006); idem., "Post-Modern Challenges for Modern Warriors," published in *Shrivenham Papers* Number 5, December 2007. Though using different words, Frank Kitson, *Low Intensity Operations* (London, Faber and Faber, 1971) focuses on the same issue, as does Gen Rupert Smith, *The Utility of Force: The Art of War in the Modern World* (London, Penguin, 2005).

[27] Marine Corps Warfighting Publication 5-1: *Marine Corps Planning Process* (Washington, DC: Department of the Navy, 24 Sept 2001), 1-2. Henceforth MCWP 5-1.

[28] Ibid.

[29] MCDP 1, 9.

[30] John C. Berry, Jr, "An Assessment of the MCPP," *Thoughts on the Operational Art* (Marine Corps Combat Development Command: Marine Corps Warfighting Lab, Oct 2006), 25-6.

[31] MCDP 5, 15.

[32] For shared situational awareness as a central function *and* goal of planning as a process, see Joint Publication 5-0: *Joint Operation Planning* (Washington, DC: Joint Chiefs of Staff, 20 December 2006), I-13, 14.

[33] MCDP 5, 17

[34] MCDP 1, 7

[35] MCDP 1, 18.

[36] MCDP 5, 4.

[37] John C. Berry, Jr, "An Assessment of the MCPP," 27; Steven A. Hardesty, "Re-

thinking the Marine Corps Planning Process: Campaign Design for the Long War," in *Thoughts on the Operational Art*, 65.

[38] MCDP 5, 24.

[39] Ibid.

[40] MCDP 5, 4, 14.

[41] Rupert Smith, towards the end of *The Utility of Force*, laments this as a characteristic of today's operational approach among post-industrial militaries.

[42] MCDP 5, 25.

[43] MCDP 1, 9, 19.

[44] John C. Berry, Jr, "An Assessment of the MCPP," *Thoughts on the Operational Art,* 27.

[45] Horst Rittel and Melvin Webber, "Dilemmas in a General Theory of Planning," in N Cross (ed), *Developments in Design Methodology* (Chichester, UK: Wiley and Son, 1984), 135-144. These "wicked" problems differ from "tame" problems of, for example, mathematics.

Appendix A

Glossary

Note: words in *italics* indicate a term defined elsewhere in this glossary.

Acephalous Societies: Literally meaning "without-a-head;" groups that have no one formal designated leader.

Affective Domain: Domain of learning associated with *Bloom's Taxonomy*, whose descriptors are "receiving," "responding," "valuing," "organization," "values complex." Affective skills involve how a person relates to and interacts with what they learn, and how they come to own it both intellectually and psycho-emotionally.

Authority: the legal or popularly granted permission to exercise *power*; legitimacy in the exercise of power.

Band: Small group of people who all know each other face-to-face, and work closely together for a unified purpose of survival. Mutual dependence requires cooperation, militating against hierarchies beyond those of age and gender. Egalitarian social structure tends to leadership selection on basis of skill.

Belief: A certainty, learned through inherited group experiences and practices, about the substance and meaning of phenomena and human activity. An individual's beliefs are relatively immune to influence by personal experiences and the environment.

Bloom's Taxonomy of Learning: Taxonomy of educational objectives and skills. Proposed in 1956 by Benjamin Bloom, an educational psychologist at the University of Chicago. Bloom's Taxonomy divides educational objectives into *Affective*, *Psychomotor*, and *Cognitive domains*. Taxonomy is hierarchical. Taxonomy's intent is to motivate educators to focus on all three domains, creating holistic form of education.

Chiefdom: *Political structure* with centralized chief, subordinate council of advisors and functionaries, not necessarily lineage-based. Subordinates' power is based on "billet" at least as much as on name. Chiefdoms are *stratified*, but centralize *authority*, decision making, and administrative structures.

Class: A way of *stratifying* groups of people according to their economic status and power in a society. Certain social characteristics such as the accumulation of goods or other forms of wealth; education; occupation; region of origin; lineage; and social behavior may all be indicators of class. However, since these indicators are culturally coded, class will be based on different characteristics in different societies.

Cognitive Domain: Domain of learning associated with *Bloom's Taxonomy*, whose descriptors are "knowledge," "comprehension," "application," "analysis," "synthesis," and "evaluation." Cognitive skills refer to the mental processes for gaining knowledge and comprehension, and involve thinking, recalling, and judging. At higher levels, cognition assists imagination and planning.

Collective Memory: A group's selective remembering, and belief-influenced attribution of meaning to historical events, most often with a present-oriented utility for that group.

Councils and Oligarchies: Seen in *tribes*, councils and oligarchies (rulership by an elite group) are also common in the leadership structures of *states* and *chiefdoms*. In a council no one individual bears the right to make a final decision or to implement a course of action. The decision must be considered and shared among all the leaders. May share characteristics in common with *acephalous* or *episodic leadership* systems as per interaction with external entities. However, councils and oligarchies possess permanency, uncontested authority, and regularized decision-making mechanisms.

Culture: The shared *world view* and social structures of a group of people that influence a person's and a group's actions and choices.

Culture Group: A group of people whose common *world view* unites them in a system of social structures and shared behaviors.

Culture Operator: Engages in military functions at the tactical, operational, and strategic level by continually re-reading the changing cultural and human aspects of the battlespace as they impact military operations; by tracking the dynamic interaction among the Operational Culture Dimensions of environment, economy, social structure, political structure, and beliefs and symbols; and by considering the impact of Marine operations as a new physical condition of human existence for indigenous people in the area of operations, influencing local behaviors and attitudes.

Culture Operations: Operations which include cultural and human factors in planning; consider possible responses to Marine actions during execution, while evaluating cultural outcomes of tactical and operational measures; weigh cultural outcomes of operational measures against mission objectives; develop innovative courses of action that allow commanders avenues of action previously unrecognized. Culture Operations emphasize planning and execution in order to create conditions facilitating indigenous conduct commensurate with tactical or operational goals, yet recognize that Marine actions are merely one among many factors influencing human conditions in the battlespace, and are unlikely to craft or determine people's behaviors.

Dictators and Strongmen: Obtain power through coercive force, and remain in power as long as they are backed by that force, which possesses the means to intimidate and coerce others in the community. Power can last only through the military and police sectors of a society, while eliminating or coopting any other potential nodes of competition. A strongman's power lacks legitimate investment of authority on the part of the people or *state.*

Distributed Operations: A form of maneuver warfare where small, highly capable units across a large area create an advantage over an adversary through deliberate use of separation and coordinated, tactical actions. Units use close combat or supporting arms to disrupt the enemy's access to key terrain and avenues of approach. Focuses on energetic training of small units and more robust communications and tactical mobility assets for those smaller units. Greater focus also placed on language and cultural training.

Economy of a Culture: Specific system for obtaining, producing and distributing the items that people need or want to survive in their society (food, water, cars, houses etc.).

Egalitarianism: Resources, *power*, and decision-making are not concentrated in the hands of any specific individual or sub-group, but are spread relatively evenly across members of the group.

Elected and Selected Leadership: Voting for leaders is extremely recent; selecting a leader on the basis of skill and experience has existed for centuries. Elected and selected leaders remain in a position of power based on effectiveness in carrying out their work—hence accountability on the part of elected and selected leaders, who are constrained by popular preferences in their decision-making.

Emotional intelligence: Connotes capabilities not usually taught in school or trained through task performance, but which develop through a cascading process of experience, reflection, and self-knowledge. Includes following traits: emotional self-awareness; independence; empathy; inter-personal relationships; stress tolerance; impulse control; flexibility; problem solving.

Episodic leaders: Found in all political structures, whether *bands, tribes, chiefdoms* or *states*. Arise for specific role, and have no formal official control or power other than that which is given to them for undertaking the specific task. Once the goal has been achieved, episodic leaders step down to continue on with their normal routine.

Ethnicity: Identification of individual with a unique subgroup in a society, distinguished by specific behaviors, characteristics, and social symbols. Include a language specific to the group; symbols reflecting group membership or carrying hidden meaning; unique traditions, rituals and holidays; dress unique to the group; shared sense, or memory of history—often enshrined in mythical stories or folk tales; attachment to a place or region that holds symbolic meaning.

Folklore: A group's collection of stories, sayings, and narratives of *history* passed down through the generations. Each generation receives this inheritance, imbues it with new meaning, and adds new narratives based on new collective experiences.

Formal Economy: Economic interactions and exchanges that are regulated, taxed, tracked and measured by a state government.

Formal Leaders: Receive official recognition by the *political structure* and community, often with formal titles, and may wear symbols indicating *status* within the community. Typically have special offices and receive various legitimate financial and other regularized benefits. Can claim *power* that comes with their *position* regardless of their knowledge, background or skills.

Great Tradition: The formal, written canonic version of a religion.

Hereditary Leadership: Leadership inherited along family lines. By default, concentrates power in the same *ethnic*, religious, socio-economic, and regional groups. Hereditary leadership discourages political mobility by reinforcing *social structures* and *stratification*: the same groups continue to be included in decision making, while others are excluded generation after generation.

History: What happened in the past.

Human Terrain: Those cultural aspects of the battlespace that, due to their static nature, can be visually represented on a geographic map. Human terrain is static with respect to change over time; rigid with respect to fluid human relationships; and limited to representing behavior in only two dimensions.

Icons: Individuals who become larger than life and symbolize many of the positive (or negative) values of a society. Prominent in both *folktales* and remembered histories. Originally were physical objects felt to represent a deity or holy person. In Orthodox Christian communities, "ikonos" were objects with representations of the Virgin Mary, Jesus Christ, or other saints, through which people channeled prayer.

Imagined Memory: The selective remembering, and belief-influenced attribution of meaning to *historical* events.

Informal Economy: Economic interactions that are not recognized, regulated, controlled, or taxed by a state government.

Informal Leaders: May not carry titles and symbols indicating their status, and spaces may not obviously indicate a person of power to an outsider. Or, may hold a formal position significantly lower or different from actual *authority* and *power*. Despite lack of official trappings, informal leaders may carry more *power* or influence over the community than *formal leaders*. Informal leaders gain status through skills in working with people and undertaking leadership tasks. Alternatively, can gain influence in a formal system because community already perceives their *status*, prestige, or skill. Typically in order to "get anything done," one needs to have the approval and support of informal leader(s).

Irregular Warfare: Use of violent or non-violent means to maintain or undermine the credibility and/or legitimacy of a political *authority* by the application of indirect approaches and non-conventional means to defeat an enemy by subversion, attrition, or exhaustion rather than through direct military confrontation. May employ the full range of military and other capabilities to seek asymmetric advantages, in order to erode an adversary's power, influence, and will.

Little Tradition: The local, informal, daily practices of a religion, which vary from region to region and even community to community.

Mores: Implicit or explicit rules regarding permissible or forbidden behavior. In contrast to *norms*, violations of a cultural more are usually accompanied by serious repercussions.

Nation: Idea of a commonality of identity and destiny among a group of people, often with political implications. Based on shared traits such as (but not limited to or requiring) language, *ethnicity*, history, religion, memory of past, aspirations for future. Nationalism is the ideology motivating political movements on the basis of these traits.

Non-Verbal Symbolic Communication: Use of body language, dress, physical positioning, design of structures, in order to communicate *status*, desire, mood, etc; usually based on symbols, values, ideals, images accepted or known among a particular culture group. All non-word-based intentional communication.

Operational Culture: Operationally relevant behavior, relationships and perceptions of indigenous security forces against or with whom Marines operate; civilian populations among whom Marines operate; indigenous communities or groups whom Marines wish to influence; international partners in coalition operations.

Dimensions influencing operationally-relevant behavior, conduct, and attitudes. These Operational Culture Dimensions involve the physical environment; the economy of a culture; social structures; political structures; and the beliefs and symbols of a culture group. These dimensions emerge from three major models of cultural analysis: the ecological, social structure, and symbolic models.

Historical trends that influence the interaction among those cultural dimensions.

Capability to successfully plan and execute across the operational spectrum, including humanitarian assistance and disaster relief; pre-hostility; shaping operations; successive campaign phases; and post hostilities, to include reconstruction and stabilization, as well as peace making/keeping.

Operational Culture Learning: In pre-deployment phase, scaled to rank and billet and focused on the AO as aligned with mission goals: study of a specific AO's culture, to include expressed behaviors and attitudes, as well as the interaction among Operational Culture Dimensions which produce these behaviors and attitudes; training in operation, as well as billet-focused language domains; attendance of command-provided distance learning, function-focused face-to-face seminars, field exercises, and situational training; and self-study, to ensure learning in the *cognitive, psycho-motor* and *affective domains.*

In PME, reflecting the responsibilities of Marines at the completion of each stage: study of the fundamental concepts and Dimensions of culture "in generica;" development of skills to function successfully in various geographical and diverse human environments; examination of human, print, and electronic resources for learning about *operational culture*; exploration of the historical role of the Five Operational Culture Dimensions in the battlespace through study of past areas of operation; and introduction to the application of concepts and skills to the current operating environment.

In the career continuum, appropriate to MOS, phase of career, and leadership responsibilities: study of emerging operating environments; maintenance of capabilities with respect to regions of past or likely future areas of operation; monitoring service- and DOD-provided training and educational resources for culture learning; fostering within units and commands continued culture learning and an atmosphere supportive of individual Marines' study of foreign cultures; recording culture-relevant observations about areas of deployment.

Physical Environment: The physical, non-human features of an environment, to include physical terrain, climate, and man-made structures.

Physical Symbols: Any physical object that holds a symbolic meaning greater than its practical utility for the people in a group. Frequently physical symbols, both natural and man-made, mark out status and/or identity. Clothing, headgear, material adornment, and symbolic objects may all indicate one's social membership and identity. So do physiognomic elements such as hair, piercing, or scars. Symbols provide information about *status*, roles, etc., to members within a group, and indicate who is in and who is out of a group.

Political Structure: The way that power and leadership are apportioned to people according to the *social structure* of the society.

Position: The symbolic place one holds relative to others in a social structure.

Power: The ability to "control" or "influence" the behavior of individuals or groups of people.

Psycho-Motor Domain: Domain of learning associated with *Bloom's Taxonomy*, whose descriptors are "imitation," "manipulation," "precision," "articulation," naturalization." Psycho-Motor skills involve movement of the body in order to perform actions based upon input from the brain.

Regime: Ruling government, not to be confused with *state* or *nation*.

Religious Membership: Being part of a group of people that considers itself united by religious faith.

Rituals: Often characterized by the notion that the actions in the ritual themselves must be performed a special way to be valid. In religious rituals, the necessity to perform every step of every ritual correctly is related to the conviction that the spiritual soundness—or metaphysical validity—will only be preserved if each step of that ritual is performed correctly. In these cases, imperfect performance violates holiness itself.

Social Environment: Those features, processes, and interactions in an AO that structure human relationships and guide interactions among people.

Social Norms: Cultural expectations about how one ought to behave in a given situation. Social guidelines that most people follow most of the time. However, norms are not rigid; people may accidentally or deliberately ignore with only minor repercussions.

Social Structure: How people organize their political, economic, and social relationships.

State: Characterized by centralized *authority* and control, defined territorial boundaries, legitimate power through militaries, police or other militias, and control of access to resources. Often possess a highly centralized government over which there is no contention, and defined territorial parameters—unlike *bands, tribes,* or *chief-doms.* States possess regularized security structures loyal to the state itself, and not something outside the state's boundaries. The security forces are major legitimizing tool for the state to effectively assure its power over the people.

Status: the meaning and value accorded by members of a *social structure* to a particular person occupying a specific *position* in that social structure. That meaning and value derives from social attitudes to the position itself; from specific individual qualities of the person; or from both.

Stratification: Resources, *power*, and decision-making are limited to certain categories of people within a community, based upon *status*, entitlement, and rank. Society is thus multi-leveled, with groups enfranchised or disenfranchised according to their place in lower or higher social levels.

Symbols: A unique characteristic of man, facilitating communication. Symbols can be physical such as a flag; verbal as in a spoken language; or behavioral as in non-verbal communication. Symbols have culturally dependent meanings.

Syncretism: The amalgamation of local cultural traditions, indigenous religious beliefs, and formal religious systems to create a synthesis of beliefs and traditions, which, when compared to the "proper" religion of scriptures, clergy, and capital-city seminaries, looks "strange" and "incompatible" with the "real" religion.

Taboos: Activities or uses of physical objects that are explicitly forbidden. Generally based on religious notions of permissible and impermissible. In contrast to *mores*, taboos are rarely about "what you should/must do," but are about "what you should/must *not* do." Broken taboos may not always carry the heavy repercussions of violations of a social *more*. This may be attributed to the notion that God or the powers that be will strike down the offender, so that others in the society need not enforce the requirement.

Tribe: A specific type of kinship group based on a shared real (or fictive) ancestor and structured around a clear corporate identity. This corporate identity creates a sense of unity and belonging among members of the tribe; and structures relationships among tribal members. Individuals within the tribe are assigned specific roles, positions and power according to their place within the lineage; some tribal lineages being considered of lower or higher status than others. Tribes have a *formal leader* or *council* of leaders, designated to speak for the group, who are selected, at least in part, on the basis of their inherited position within the tribe. It is important to note that many kinship groups are **not** tribes.

World View: The way that people perceive and think about their world.

Appendix B

Culture Operator's Questions

Below are the "Culture Operator's Questions" for each component of the Five Operational Culture Dimensions.

Dimension 1: Environment

Water

- What are the cultural rules about water's use?

- What roles are expected of Marine personnel with respect to water use and provision?

- What is the relationship between water use and ritual?

- What is the symbolic significance of water?

- Who customarily has what jobs, roles, and functions with respect to water?

- Who, in the AO, has customarily controlled access to water, and how have they used that for power, influence, etc.?

- What is the scarcity of water in relation to intensity to use?

- What kinds of operational considerations are influenced by water, or override cultural aspects of water as a physical resource?

Land

- What are the symbolic meanings of certain sub-districts in the AO, and how do groups within the AO view this symbolism differently?

- What are particular land formations that are visually striking, with local significance?

- What land in the AO is/is not appropriate for certain groups of people to use?

- Who, locally, has legitimate ability to determine outsiders' access to land?

- What are the local conventions of private, communal, and state ownership/use of land?

- What is the relationship between the political map's national/regional boundaries, and what people living in the AO see as the boundaries that matter, in political, economic, genealogical, and security terms?

- What are the geographic area's principles of division, and relationship between dividing lines and access to both tangible and symbolic resources?

Food

- What are the local staples, and what is the required labor to grow, prepare and serve them?

- What kinds of locally-accepted foods are considered strange, dangerous, or not even food to less-traveled Marines?

- What foods—eaten either by U.S. personnel or by local people—are so out of place as to raise concerns about health or sanitation?

- What foods are served by whom, to indicate the status of server or guest?

- How do Marine operations or logistics impact the ability of local people to obtain essential foodstuffs?

- What foods have which kinds of ritual significance?

- What are the time- or calendar-related roles of what kinds of foods?

- Which foods are strategic commodities, inasmuch as controlling access to them influences one's coercive or political power?

- What, in local terms, is considered food sufficiency, food scarcity, and the proper role of external forces in providing food?

Materials for Shelter

- How do structures fit the geographic, climatic, and physical aspects of the environment?

- What do the internal and external appearances and materials used in structures communicate about building purpose, occupant status, etc.?

- What materials for building, repair, and maintenance are local to the AO?

- What are the central tactical implications of building styles, neighborhood layout, etc?

Climate and Seasons

- How does the climate influence local attitudes to, and capabilities for work, business, and combat?

- What is the relationship between climate and season, on the one hand, and battle rhythm and operational tempo, on the other?

- What, in local terms, passes for good weather, bad weather, etc?

Fuel and Power

- What are the locally found, or locally produced sources of power and fuel?

- What is the relationship between local elites and access to/provision of fuel and power?

- How does the larger government authority provide, or control access to power?

- What do local people expect of outside forces in terms of power/fuel provision and protection?

- What are local work-arounds to deal with shortages of power and fuel, and how do Marine operations impact them?

- What local issues regarding power and fuel are overshadowed by more pressing operational considerations?

Dimension 2: Economy

Informal Economy

- What categories of people are active in the AO's informal economy?

- On what commodities/services does the informal economy focus?

- What is the relationship between the informal economy, on the one hand, and unregulated movement of people, crime, and violence, on the other?

- How will Marine operations impact the informal economy and the people in it?

- How will the Marine impact on the informal economy influence attitudes of certain sectors of the population to the Marine presence?

- How does the formal economy rely upon the informal economy, and what abuses of the AO's population does this cause?

- What opportunities exist for the AO's population, based on the formal economy's relationship with informal economic practices?

- What are formal/informal economic actors' expectations of the state or over-arching political-military authority, with respect to involvement in or disregard for economic activity?

- What is considered an "illegal" good or service in the AO, on what basis?

- What goods/services are legal, but culturally frowned upon? Who deals in these goods/services?

- How will Marine expenditure in the local informal economy, or employment of local informal economic actors, influence the socio-economic balance of power in the AO?

Economy as a Network of Exchange

- How are important physical resources (food, clothing, shelter, cars etc.) obtained by local peoples?

- How do people gain access to critical services such as medical care, transportation, or education?

- Would a specific operational plan improve or block access to critical goods and services?

- What is the degree of (in)equity in the distribution of goods and services among the population?

- Who seems to control the distribution of goods and services, and how? Would a planned operation change this distribution pattern?

- Along with or instead of money, what do local peoples rely on to obtain and exchange goods in the region?

- If money is not the primary economic system, how could the operational plan be adjusted to use the existing alternate economic systems effectively?

Economy as a Way of Structuring Social Relationships

- What are the main economic systems in place in the region (pastoralism, agriculture, industrial production—all three may be present simultaneously)?

- What are the economic rhythms of the community (migration seasons, planting and harvesting, market day, work hours)?

- What are the important features of the environment that determine the economy of the AO?

- How is wealth distributed? Does wealth seem to be concentrated in the hands of certain individuals or groups? On what basis?

- How do local economic structures reflect the relationship of the group to the larger political and state system?

Dimension 3: Social Structure

Age

- At what age is someone considered a child or adult?

- What specific ceremonies mark the transition to adulthood?

- Which new social privileges are granted to men and women when they pass these manhood or womanhood rituals?

- What are locally accepted or expected economic roles for what U.S. society considers children?

- How should Marines prepare to respond to children that act as soldiers in militaries or insurgencies, or participate in violent activities against U.S. forces?

- What special status or roles are accorded to the elderly?

- Is there an age grading system that stratifies people according to their age and stage in the life cycle? And if so, what rights, roles and duties do people have at each stage?

Gender

- What work, roles, activities and spaces are assigned predominantly to men and women?

- Who undertakes which tasks and where?

- In what ways must operational plans be adjusted to account for the different work, roles, and spaces assigned to men and women?

- What roles do women play in local militaries and insurgencies? Do they engage in armed combat?

- If women are not visibly observable, what roles and tasks do they undertake 'behind the scenes'?

- How can operational plans and assignment of manpower include gender to maximize effectiveness of the unit?

Kinship, Clan, and Tribal Membership

- How are land, water, or access to certain goods and resources concentrated in the hands of specific kin groups or tribes?

- How will our operations in the region support certain kin groups and enhance their power; or conversely undermine these groups?

- What are the possible outcomes of an operation that will challenge the power or control of resources by certain kin groups in the region (war, insurgency, increased stability, greater/lesser access to important goods and services)?

- How does a Marine's choice of local points of contact interact with or disturb local kin relationships, thus influencing the degree of success of Marine initiatives?

Class

- How is class defined in the AO: on the basis of wealth, education, region of origin, inheritance, or other factors?

- What are the privileges (economic, political, social, religious) of members of the upper class?

- How is access to essential resources for survival (food, shelter, clothing, water) determined by class?

- How does the concentration of wealth (through corruption, graft, or legitimate means) in the hands of an elite upper class relate to resource or power access?

- In creating a plan to support lower class groups, will funds and resources have to pass through the hands of the upper class first (and consequently disappear)?

- What is the reality of upward mobility in the AO's class system, and what do local people consider to be their potential for in-system upward mobility?

- How will Marine measures that influence different groups' social mobility be viewed by those groups, or by other, competing groups?

Ethnicity

- What is the relationship between particular ethnic groups and control of professions or positions of formal or informal power?

- How do groups that are barred from these positions of power challenge the system (breeding grounds for insurgents, theft and bribery, civil war)?

- What are local assumptions about U.S. and western biases and partisanship with respect to ethnic group struggles?

- How will a Marine alliance or dealings with a particular ethnic group affect those in power?

- What are the possible reactions of those groups that are ignored?

Religious Membership

- How do people define and express their religious membership in the region?

- What roles and status do the various religious groups or sects hold in the larger society?

- What is the meaning of geography for religious groups in an AO?

- What effects would a planned Marine operation in the region have upon the power, status and access to critical resources of the various religious groups or sects?

- How will the Marine operations influence indigenous peoples' views of Marine or U.S. biases towards different religious groups of the social structure?

Dimension 4: Political Structures

Leadership

- How is decision-making organized, and who gets to make make decisions?

- What are the principles and processes governing policy deliberations and decision-making?

- Who do leaders have to consult, and to whom must they answer?

- How is leadership obtained and passed on (by election, inheritance, demonstration of skill, membership in a certain age or social group, by force)?

- Who are formal leaders and what symbols indicate status?

- If one needs to get something done, to whom do people turn?

- What is relationship between the formal and informal leader?

Conflicts over Power

- What are the most important cultural characteristics that determine one's position and power in the community (age, class, gender, tribal identity, ethnicity, religion)?

- What is the degree of polarization in the region with respect to religious/ethnic/tribal identities?

- What is the amount of flexibility and interaction between religious/ethnic/tribal groups?

- Which groups hold power, to what degree of concentration?

- Which groups are excluded, and along which axes?

- What is their degree of consciousness of exclusion?

- How possible do group leaders think it is to challenge the system?

- How do marginalized and losing groups gain access to valued goods and resources and opportunities?

- How will alliance with one group affect Marine relationships with the other groups?

Dimension 5: Beliefs and Icons

History, Imagined Memory and Folklore

- What are the pivotal historical stories that all people in the community share?

- How do different groups in the AO give different significance to the same historical stories?

- What are the daily sayings and folktales that everyone refers to in common conversation?

- How are these remembered histories, folktales and sayings used to emphasize or teach important values and ideals?

- How are these histories, folktales and sayings used to support propaganda for or against Marine and U.S. activities in the region?

Icons

- Who are the local heroes? What important qualities do these heroes embody?

- Who are the local villains? Why are they villainous (what makes them evil)?

- Are the heroes or villains compared to Marines or Americans?

- What do the comparisons illustrate about local attitudes towards the U.S. and the military?

Symbols and Communication

- What physical symbols (clothing, headdress, scarification, insignia) indicate membership or status in the ethnic, religious and social groups of the region?

- What physical and written symbols (signs, graffiti, fences, spiritual markers) are important to be able to recognize in order to navigate and understand what is happening in the region?

- What words or phrases are essential for basic communication with local people?

- What non-verbal behaviors may be misinterpreted by local people? Which non-verbal behaviors (such as seating patterns) are important to understand in meetings and negotiations?

Rituals and Ceremonies

- What behaviors and actions are important in the ritual or ceremony and what does this reveal about cultural ideals and values?

- Who participates in the ritual, and what roles do the participants play?

- What does presence of participants, or the nature of their participation, say about their membership and status in the group?

- What does the public performance of the ritual communicate to outsiders?

- How is this performance potentially a politically charged statement about the group's status and rights within the larger society?

- What activities, not related to the ritual or ceremony itself, occur at ceremonial gatherings, due to the social status of the participants?

Norms, Mores, and Taboos

- What food and behavioral taboos exist in the region?

- What norms should Marines, even though they are foreign to the AO, observe?

- What underlying allegiances or codes of honor could influence the success of an operation?

- What activities in the area are considered serious violations of social mores and could carry serious punishments including death?

- What beliefs or assumptions exist locally about American practices as regards local norms, mores, and taboos?

- What might the local people think (or have been propagandized to think) Marines are likely to disregard in terms of local norms, mores, and taboos?

Religious Beliefs

- Who is the actual leader of the local religious community?

- How do religious leaders relate to the educated elite vs. popular groups, etc.?

- What is the basis of authority for a "religious" leader in the AO: book learning, lineage, charisma, etc.?

- What are the actual (versus theoretical/textual) religious practices in the specific AO where the Marine operates?

- How do local practices of a religion the Marine has encountered elsewhere differ from what that Marine thinks the religion is "supposed" to look like?

- What power and role, if any, does the formal religious system play in local peoples' daily lives?

- What conflicts or disagreements exist between the formal religious system and the local religious practices of the AO?

- How prominent is "religion" as an explanatory factor for people in current events, and in reference to history, or historical trajectories?

- What is "the way the world is supposed to be" according to locally-held religious beliefs, and how does the Marine presence impact that?

Bibliography

Abaza, Khairi. "Sharm al-Shaykh Bombings: The Egyptian Context." The Washington Institute for Near East Policy, August 12, 2005: http://www.washingtoninstitute.org/templateC05.php?CID=2355

Abazov, Rafis. *Culture and Customs of the Central Asian Republics.* Westport, CT: Greenwood Press, 2007.

Abbe, Allison. "Developing Cross-Cultural Competence in Military Leaders." *Pedagogy for the Long War: Teaching Irregular Warfare – Conference Proceedings* Barak Salmoni, comp. and ed. Marine Corps Training and Education Command, 2008.71-72.

_____, with Lisa M.V. Gulick, and Jeffrey L. Herman. *Cross Cultural Competence in Army Leaders: A Conceptual and Empirical Foundation.* US Army Research Institute for the Behavioral and Social Sciences, Study Report 2008-01, October 2007.

Abu Lughod, Janet. "Migrant Adjustment to City Life: the Egyptian Case." In *Arab Society in Transition*, edited by I. Saad and N.S. Hopkins. Cairo: The American University, 1977: 391-405.

Adler, Margot. "Chinese Restaurant Workers Protest Lost Wages." (Audio Recording) *Morning Edition.* National Public Radio, May 8, 2007.http://www.npr.org/templates/story/story.php?storyId=10089965.

"'Academic Embeds': Scholars Advise Troops Abroad," *Talk of the Nation* (National Public Radio), 9 October 2007: http://www.npr.org/templates/story/story.php?storyId=15124054

Alexander, MAJ Clinton D. "Racial Diversity in the Marine Corps." MMS Paper, Marine Corps Command and Staff College, 2008.

Alexander, M., M. Evans, J.F.V. Keiger, eds. *The Algerian War and the French Army, 1954-62: Experiences, Images, and Testimonies* New York: Palgrave Macmillan, 2002.

Alfassa, S. "Is Uncle Sam Confused," *The Jewish Press*, 24 Oct 05.

Allawi, Ali. *The Occupation of Iraq: Winning the War, Losing the Peace.* New Haven, CT: Yale University Press, 2007.

Amos, Deborah. "Iraqi Women Fight for Voice in New Iraq." (Audio Recording) Morning Edition. *National Public Radio*, 3 July 2003. www.npr.org/templates/story/story.php?storyId=1319135

Anderson, Lorin W. and David R. Krathwohl. *A Taxonomy for Learning, Teaching, and Assessing: A Revision of Bloom's Taxonomy of Educational Objectives.* Boston, MA: Allyn and Bacon, 2000.

Anonymous CAG Team Leader. Interview by Barak Salmoni. Naval Post-Graduate School, September 2004.

Arcilla, Jose S. *An Introduction to Philippine History.* Manila: Ateneo de Manila University Press, 2003.

Arnold, Kris A. *PMESII and The Non-State Actor: Questioning the Relevance.* Ft Leavenworth, KS: Army Command and General Staff College – School of Advanced Military Studies, May 2006.

Avruch, Kevin. *Culture & Conflict Resolution.* Washington, D.C.: United States Institute of Peace Press, 1998.

Bailey, Garrick and James Peoples. *Introduction to Cultural Anthropology.* Belmont, CA: Wadsworth Publishing Co., 1999.

Banlaoi, Rommel C. "Maritime Terrorism in Southeast Asia." *Naval War College Review* 58:4 (Autumn 2005).

Baram, Amatzia. "Neo-Tribalism in Iraq: Saddam Hussein's Tribal Policies 1991-96." *International Journal of Middle East Studies* 29:1 (Feb. 1997): 1-31.

_____. *Culture, History and Ideology in the Formation of Ba'thist Iraq,* 1968-89. New York: Palgrave Macmillan, 1991.

Barnard, Alan. *History and Theory in Anthropology.* Cambridge, UK: Cambridge University Press, 2000.

Barnouw, Victor. *Culture and Personality.* Homewood, Ill.: Dorsey Press, 1963.

Baxter, P.T.W., and U. Almagor, eds. *Age, Generation, and Time: Some Features of East African Age Organizations*. New York: St Martin's Press, 1978.

Ben Ari, Eyal. *Mastering Soldiers: Conflict, Emotions, and the Enemy in an Israeli Military Unit*. London: Berghan Books, 1995.

Bendersky, Joseph W. *The "Jewish Threat": Anti-Semitic Politics of the U.S. Army*. New York: Basic Books, 2001.

Benedict, Ruth. *Patterns of Culture*. Boston: Houghton Mifflin, 1934.

Ben Eliezer, Uri. *The Making of Israeli Militarism*. Bloomington: Indiana University Press, 1998.

Benvenisti, Meron. *Sacred Landscape: The Buried History of the Holy Land since 1948*. Berkeley, CA: University of California Press, 2002.

Berkes, Niyazi. *The Development of Secularism in Turkey*. London: Routledge, [1964] 1999.

Berry, John C., Jr, "An Assessment of the MCPP," *Thoughts on the Operational Art*. Marine Corps Combat Development Command: Marine Corps Warfighting Lab, Oct 2006.

Birand, Mehmet Ali. *Shirts of Steel: Anatomy of the Turkish Officer Corps*. London: IB Taurus, 1991.

Bloom, Benjamin. *Taxonomy of Educational Objectives, Handbook 1: Cognitive Domain*. Boston, MA: Addison Wesley Publishing, 1956.

Bloom, Benjamin S., David R. Krathwohl, and Bertram B. Masia. *Taxonomy of Educational Objectives, Book 2: Affective Domain*. New York: Longman, 1999.

BN CO from 1st MarDiv. Interviewed by Barak Salmoni. OIF II, TCMEF, April 2005.

Boas, F. *Race, Language and Culture*. New York: Macmillan, 1940.

Boktor, Amir. *School and Society in the Valley of the Nile.* Cairo: Elias' Modern Press, 1936.

_____. *The Development and Expansion of Education in the U.A.R.* Cairo: AUC Press, 1963.

Braude, Benjamin and Bernard Lewis. *Christians and Jews in the Ottoman Empire: The Functioning of a Plural Society.* Princeton, NJ: Holmes and Meier, 1982.

Burnett, John. "Violence Surges Along the U.S.-Mexico Border." (Audio Recording) *All Things Considered.* National Public Radio, February 12, 2006: http://www.npr.org/templates/story/story.php?storyId=5203014.

Burr, J. Millard and Collins, Robert O. *Darfur: The Long Road to Disaster.* Princeton, NJ: Markus Wiener Publishers, 2006.

Burrowes, Robert D. "Oil Strike and Leadership Struggle in South Yemen: 1986 and Beyond." *The Middle East Journal* 43:3 (Summer 1989).

Camp, Dick. *Leatherneck Legends: Conversations with the Marine Corps' Old Breed.* Osceola, WI: Zenith Press, 2006.

Cassidy, R. M. "The British Army and Counter-insurgency: The Salience of Military Culture," *Military Review.* May 2005.

Cassirer, Ernst. *An Essay on Man.* New Haven, CT: Yale University Press, 1944.

Catagnus, Edison, Keeling, and Moon. "Infantry Squad Tactics." *Marine Corps Gazette.* September, 2005.

Caton, Stephen C. *Yemen Chronicle: An Anthropology of War and Mediation.* New York: Hill and Wang, 2005.

Center for Advanced Operational Culture Learning. *Afghanistan: Operational Culture for Deploying Personnel.* Quantico, VA: Marine Corps Training and Education Command, 2006.

Chagnon, Napoleon A. *The Yanomamo.* Belmont, CA: Wadsworth Publishing, 1996.

Chaitin, Julia. *Inside-Out: Personal and Collective Life in Israel and the Kibbutz.* Lanham, MD: University Press of America, 2007.

Clinton Robert N. and Carole E. Goldberg. *American Indian Law: Native Nations and the Federal System,* 5th ed. Matthew Bender: 2007.

Coronel, Sheila, Yvonne Chua, Luz Rimban, and Booma Cruz. "The Rulemakers: How the Wealthy and Well-Born Dominate Congress." *Philippine Centre for Investigative Journalism,* 2004.

Cratty, Bryant J. *Psycho-Motor Behavior in Education and Sport.* Charles C. Thomas Pub Ltd, 1974.

Crews, Robert D. and Amin Tarzi, eds. *The Taliban and the Crisis of Afghanistan.* Cambridge, MA: Harvard University Press, 2008.

Dahl, Stephan. *Intercultural Skills for Business.* London: ECE, 1998.

D'Atkine, Norville. "Why Arabs Lose Wars," *Middle East Review of International Affairs* 4:1 (2000).

Dawood, Hosham. "The 'State-ization' of the Tribe and the Tribalization of the State: the Case of Iraq." *In Tribes and Power: Nationalism and Ethnicity in the Middle East.* Edited by Faleh Abdul-Jabar and Hosham Dawood. London: Saqi, 2003.

"Debrief: 2nd BN 7th MAR CAP Plt." TCMEF. 29 Palms MCB, Nov 2004.

"Debrief of Director, ISF Cell of 1st Marine Division," TCMEF. Naval Postgraduate School, January 26, 2005.

"Final Report: I MEF Headquarters Group". Interview by Dr. Barak Salmoni and CWO2 Rob Markiewicz. Al Anbar Province, Iraq. October, 2006.

"Debrief of BN-ETT to Afghan National Army, May 2005," Conducted by Barak Salmoni. Center for Advanced Operational Culture Learning, May 2005.

"Debrief of Seth Moulton and Anne Gildroy." *Team Phoenix Report.* CAOCL, April 4, 2006.

Dekmejian, R. Hrair and Hovann H. Simonian,. *Troubled Waters: The Geopolitics of the Caspian Region.* London: I. B. Tauris, 2003.

Dodge, Toby. *Inventing Iraq: The Failure of Nation Building and a History Denied.* New York: Columbia University Press, 2005.

Doumani, Beshara. "Rediscovering Ottoman Palestine: Writing Palestinians into History." *Journal of Palestine Studies* 21: 2 (Winter 1992): 5-28.

Druckman, Daniel. "Social-Psychological Factors in Regional Politics." In Werner Feld and Gavin Boyd, eds. *Comparative Regional Systems: West and East Europe, North America, The Middle East and Developing Countries.* New York: Pergamon Press 1980.

Durham, Deborah. "Youth and the Social Imagination in Africa: Introduction to Parts 1 and 2." *Anthropological Quarterly* 73:3 (July 2000): 113-120.

Duthu N. Bruce and Colin Calloway, eds. *American Indians and the Law.* New York: Viking Press, 2008.

Duyar-Kienast, Umut. *The Formation of Gecekondu Settlements in Turkey:* The Case of Ankara. Lit Verlag [ABC], 2006.

Earley, P.C. "Redefining Interactions Across Cultures and Organizations: Moving Forward with Cultural Intelligence," *Research in Organizational Behavior* 24, 2002, 271-99.

Early, Evelyn. *Baladi Women of Cairo: Playing With an Egg and Stone.* Boulder, CO: Lynne Riener, 1993.

Eickelman, Dale F. *The Middle East and Central Asia: An Anthropological Approach.* Upper Saddle River, NJ: Prentice-Hall, Inc, 1998.

English, Allan. "The Influence of Our Military Culture on What is Taught in Our Professional Military Institutions," in *Pedagogy for the Long War: Teaching Irregular Warfare – Conference Proceedings,* edited by Barak Salmoni. Marine Corps Training and Education Command, 2008.

Entelis, John P. *Islam, Democracy, and the State in North Africa.* Bloomington, IN: Indiana University Press, 1997.

Estes, Kenneth W. *The Marine Officer's Guide*, 6th edition. Annapolis, MD: US Naval Institute Press, 1996.

_____, and R.D. Heinl. *Handbook for Marine NCOs*, 4th edition. Annapolis, MD: Naval Institute Press, 1995.

Evans-Pritchard, E.E. Kinship and Marriage Among the Nuer. Oxford: Clarendon, 1951.

_____. *The Nuer.* Oxford: Clarendon, 1940.

Faris, Stephan. "The Real Roots of Darfur." *The Atlantic Monthly*, April 2007.

Ferraro, Gary. *Cultural Anthropology: An Applied Perspective.* 5th edition. Belmont CA: Thomson Wadsworth, 2004.

Flint, Julie and de Waal, Alex. *Darfur: A Short History of a Long War.* London: Zed Books, 2006.

Folsom, Seth. *The Highway War: A Marine Company Commander in Iraq.* Dulles, VA: Potomac Books, 2006.

Frake, Charles O. "Abu Sayyaf: Displays of Violence and the Proliferation of Contested Identities among Philippine Muslims." *American Anthropologist* 100:1 (March 1998): 41-54.

Frank, Richard. *Guadalcanal: The Definitive Account of the Landmark Battle.* New York: Random House, 1990.

Galula, David. *Counterinsurgency Warfare: Theory and Practice.* Westport, Conn: Praeger Security International, [1964] 2006.

Gardner, H. *Frames of Mind: The Theory of Multiple Intelligences.* New York: Basic Books, 1983.

Garrick, Bailey and James Peoples. *Introduction to Cultural Anthropology.* Belmont CA: Wadsworth, 1999.

Geertz, Clifford. *Islam Observed: Religious Development in Morocco and Indonesia.* Chicago, Ill.: University of Chicago Press, 1968.

_____ . *Interpretation of Cultures.* New York: Basic Books, 2000.

Goldschmidt, Arthur, Amy J. Johnson and Barak A. Salmoni, eds. *Re-Envisioning Egypt, 1919-1952*. Cairo: American University in Cairo Press, 2005.

Goleman, David. *Emotional Intelligence*. New York: Bantam, 1995.

Goodenough, Ward. *Culture, Language, and Society*. Reading, MA: Addison-Wesley, 1971.

Goodson, Larry P. *Afghanistan's Endless War: State Failure, Regional Politics, and the Rise of the Taliban*. Seattle, WA: University of Washington Press, 2001.

Gorenberg, Gershon. *The End of Days: Fundamentalism and the Struggle for the Temple Mount*. London: Oxford University Press, 2002.

Griffith, Samuel B. *The Battle for Guadalcanal*. Champaign, Illinois: University of Illinois Press, 1963.

Grima, Benedicte. *Secrets from the Field: An Ethnographer's Notes from Northwestern Pakistan*. Bloomington, IN: Authorhouse, 2004.

Hajjar, Maj Remi, USA. "The Army's New TRADOC Culture Center." *US Army Military Review* (Nov-Dec 2006).

Halbwachs, Maurice. *On Collective Memory*. Introduced and edited by Lewis A. Coser. Chicago: University of Chicago Press, 1992.

Halliday, Fred. "Tunisia's Uncertain Future," *Middle East Report* No. 16. Mar-Apr 1990.

Hammes, Colonel Thomas X. (USMC). "Fourth Generation Warfare Evolves, Fifth Emerges." *US Army Military Review* May 2007: 14-23.

——————. *The Sling and the Stone: On War in the 21st Century*. Osceola, WI: Zenith Press, 2006.

Haqqani, Husain. *Pakistan: Between Mosque and Military*. Washington D.C.: Carnegie Endowment for International Peace, 2005.

_____. "Head Scarves and History: For Clumsy Secularism, Deadly Rewards." *International Herald Tribune* 22 Dec 2003.

Hardesty, Steven A. "Rethinking the Marine Corps Planning Process: Campaign Design for the Long War." *In Thoughts on the Operational Art.* Marine Corps Warfighting Lab, Oct 2006.

Harrow, Anita. *A Taxonomy of Psychomotor Domain: A Guide for Developing Behavioral Objectives.* New York: David McKay, 1972.

Hawes, Gary. "Theories of Peasant Revolution: A Critique from the Philippines." *World Politics* 42:2 (1990): 261-298.

Headquarters United States Marine Corps. "A Concept for Distributed Operations" (25 April 2005): https://www.mccdc.usmc.mil/FeatureTopics/DO/A%20Concept%20for%20Distributed%20Operations%20%20Final%20CMC%20signed%20co.pdf

Hecksher, Charles, et al. *Agents of Change: Crossing the Post-Industrial Divide.* London: Oxford University Press, 2003.

Hendry, Joy. *An Anthropology of Indirect Communication.* London: Routledge, 2001.

Heyd, Uriel. *Foundations of Turkish Nationalism: The Life and Teachings of Ziya Gokalp.* New York: Hyperion Press, 1979.

Hinnebusch, Raymond. *Syria: Revolution from Above.* London: Routledge, 2002.

Hofstede, Geert. *Cultures and Organizations: Software of the Mind.* Cambridge, UK: McGraw Hill, 1991.

_____. "National Cultures Revisited." *Behavior Science Research* 18, 1983: 285-305.

_____. *Culture's Consequences: International Differences in Work-Related Values.* Sage Publications, 1980.

Holl, Jack. "The New Washington Monument: History in the Federal Government." *The Public Historian* 7:5 (Autumn 1985), 9-20.

Holmes-Eber, Paula. *Daughters of Tunis: Women, Family, and Networks in a Muslim City.* Boulder, Co: Westview Press, 2003.

_____. "Gender in the Informal Economy: North Africa." *Encyclopedia of Women in Islamic Cultures* Vol. 4. Leiden: Brill, 2003.

_____. "Mapping Cultures and Cultural Maps: Representing and Teaching Culture in the Marine Corps." *Pedagogy for the Long War: Teaching Irregular Warfare: Conference Proceedings* Barak Salmoni, comp. and ed. Marine Corps TECOM, 2008.

_____, with Barak Salmoni, "Marine Officers' Imagined Self-Identities: Ethnographic Insights for Cultural/Psychological Training," 1st International Operational Psychology Conference, IDF Military Psychology Center, Herzliya, Israel, 01 Dec 2006.

Holmstedt, Kirsten. *Band of Sisters: Women at War in Iraq.* Stackpole Books, 2007.

Honwana, Alcinda. "Negotiating Postwar Identities: Child Soldiers in Mozambique and Angola." *In Contested Terrains and Constructed Categories: Contemporary Africa in Focus.* Edited by George Clement Bond and Nigel C. Gibson. Cambridge, MA: Westview Press, 2002.

Hookway, James. "U.S. War on Terror Shows Promise in the Philippines: Local Officers Lead, with American Aid." *Wall Street Journal.* 18 June 2007, A 1.

_____. "Terror Cells Team Up to Survive in the Philippines." *Wall Street Journal.* 22 Jan 2007, A 8.

Horne, Alistair. *A Savage War of Peace: Algeria 1954-1962.* New York: NYRB Classics, 2006.

Hussain, Zahid. *Frontline Pakistan: The Struggle With Militant Islam.* New York: Columbia University Press, 2007.

Hussmans, Ralf. "Measuring the Informal Economy: From Employment in the Informal Sector to Informal Employment." Working Paper No. 53. Geneva: International Labor Office Bureau of Statistics, Policy Integration Department. Dec 2004.

Hutchcroft, Paul D. "Colonial Masters, National Politicos and Provincial Lords: Authority and Local Autonomy in the American Philippines, 1900-1913." *The Journal of Asian Studies* 59:2 (2000): 277-306.

Imber, Colin. *The Ottoman Empire, 1300-1650: The Structure of Power.* London: Palgrave-Macmillan, 2004.

Inalcik, Halil. "Ottoman Methods of Conquest," *Studia Islamica* 2 (1954).

_____. *The Ottoman Empire in the Classical Age, 1300-1600.* London: Weidenfield and Nicolson, 1973.

International Labor Organization. "STAT Working Paper No.1 – 2002: International Labor Organization Compendium of Official Statistics on Employment in the Informal Sector." http://www.ilo.org/public/english/bureau/stat/download/compla.pdf.

Irregular Warfare Joint Operating Concept (IW JOC) Version 1.0, 11 September 2007.

Islam, Syed S. "The Islamic Independence Movements in Patani of Thailand and Mindanao of the Philippines." *Asian Survey* 38:5 (1998): 441-456.

Jabar, Faleh A. "Sheikhs and Ideologues: Deconstruction and Reconstruction of Tribes under Patrimonial Totalitarianism in Iraq, 1968-1998," In *Tribes and Power: Nationalism and Ethnicity in the Middle East.* Edited by Faleh Abdul-Jabar and Hosham Dawood. London: Saqi, 2003.

Johnson, Douglas Hamilton. *The Root Causes of Sudan's Civil Wars.* Bloomington, IN: Indiana University Press, 2003.

Johnson, Rob. *Oil, Islam, and Conflict: Central Asia Since 1945.* London: Reaktion Books, 2007.

Joseph, Suad. "The Neighborhood Street in Lebanon." In *Women in the Muslim World.* Bell, Lois. and N. Keddie (eds.), p. 541-558. Cambridge, MA: Harvard University Press, 1978.

_____. "Working Class Women's Networks in a Sectarian State: A Political Paradox." *American Ethnologist* 10:1 (1983): 1-23.

Joint Pub 5-00.1: *Doctrine for Campaign Planning*. 25 January 2002.

Joint Pub 5-00.2: *Joint Task Force Planning Guidance and Procedures*. 13 January 1999.

Joint Pub 5-0: *Joint Operation Planning*. 26 December 2006.

Kally, Elisha and Fishelson, Gideon. *Water and Peace: Water Resources and the Arab-Israeli Peace Process*. Westport, CT: Praeger Publishers, 1993.

Kedourie, Elie. *Politics in the Middle East*. London: Oxford University Press, 1993.

Keesing, Roger M. *Kin Groups and Social Structure*. New York: Holt, Rinehart and Winston, 1975.

Kelman, Herbert. "Social-Psychological Dimensions." In *Peacemaking in International Conflict: Methods and Techniques*. I. William Zartman and J. Lewis Ramussen, eds. Washington, DC.: United States Institute of Peace, 1997.

Khalaf, Issa. *Politics in Palestine: Arab Factionalism and Social Disintegration, 1939-1948*. Binghamton, NY: State University of New York Press, 1991.

Khalid, Adeeb. *Islam after Communism: Religion and Politics in Central Asia*. Berkeley, CA: University of California Press, 2007.

Khalidi, Rashid. *Palestinian Identity*. New York: Columbia University Press, 1998.

Khalidi, Tarif. "Palestinian Historiography: 1900-1948." *Journal of Palestine Studies* 10:3 (Spring 1981): 58-70.

Kihlstrom, J.F., and N. Cantor, *Social Intelligence*. London: Cambridge University Press, 2000.

Kirke, Charles. "Change and Military Culture." *Pedagogy for the Long War: Teaching Irregular Warfare – Conference Proceedings.* Marine Corps Training and Education Command, 2008, 36-37.

Kiszely, LtGen Sir John. "Post-Modern Challenges for Modern Warriors." *Shrivenham Papers* Number 5, December 2007: http://www.da.mod.uk/colleges/csrc/document-listings/monographs/shrivenham_paper_5.pdf.

——————. "Learning About Counter-Insurgency," *RUSI Journal* 152:1 (December 2006).

Kitson, Frank. *Low Intensity Operations.* London, Faber and Faber, 1971.

Kluckhohn, Clyde, Dorothea Leighton, Lucy Wales Kluckhohn, and Richard Kluckhohn. *The Navajo.* Cambridge, MA: Harvard University Press, 1992.

Kolars, John F. and William A. Mitchell. *The Euphrates River and the Southeast Anatolia Development Project.* Carbondale, IL: Southern Illinois University, 1991.

Kolb, D.A. *Experiential Learning: Experience as A Source of Learning and Development.* Engelwood Cliffs, NJ: Prentice-Hall, 1984.

Kroeber, A. L. and C. Kluckhohn. *Culture: A Critical Review of Concepts and Definitions.* Visalia CA: Vintage Press, 1954.

Krulak, General Charles C. "The Strategic Corporal: Leadership in the Three Block War." *Marines Magazine*: January 1999.

Krulak, LtGen Victor. *First to Fight: An Inside View of the U.S. Marine Corps.* Annapolis, MD: US Naval Institute Press, 1999.

Kurtz, Donald. *Political Anthropology: Paradigms and Power.* Boulder CO: Westview Press, 2001.

Kurtz, Suzanne. "Semper Chai: General Robert Magnus," *Hillel News 21* Feb 2007: http://www.hillel.org/about/news/2007/feb/magnus_2007Feb21.htm.

Labrador, Mel C. "The Philippines in 2000: In Search of a Silver Lining." *Asian Survey* 41:1 (Jan-Feb 2001): 221-229.

Lapidus, Ira. *A History of Islamic Societies*, 2nd ed. London: Cambridge University Press, 2002.

Latour, LtCol Sharon M, and LtGen Bradley C. Hosmer, USAF (ret.). "Emotional Intelligence: Implications for all United States Air Force Leaders." *Air and Space Power Journal* Winter 2002: http://www.airpower.maxwell.af.mil/airchronicles/apj/apj02/win02/latour.html.

Layton, Robert. *An Introduction To Theory in Anthropology*. Cambridge, UK: Cambridge University Press, 1997.

Levi-Strauss, C. *The Elementary Structures of Kinship*. Boston: Beacon, 1969.

Lewis, Adrian R. *The American Culture of War: The History of U.S. Military Force from World War II to Operation Iraqi Freedom*. London: Routledge, 2006.

Litvak, Meir. "A Palestinian Past: National Construction and Reconstruction." *Memory and History* 6: 2 (Fall/Winter 1994): 24-56.

Livingstone, Holly, Maria Nadjiwon-Foster, Sonya Smithers. "Emotional Intelligence and Military Leadership." *Canadian Forces Leadership Institute*, 11 March 2002: http://www.cdaacd.forces.gc.ca/CFLI/engraph/research/pdf/08.pdf.

Lodwick, Dora G. "Reviewed Work(s): Peru: When the World Turned Dark." *Teaching Sociology* 19:1 (January 1991): 117-119.

Lorcin, Patricia. *Imperial Identities: Stereotyping, Prejudice, and Race in Colonial Algeria*. London: I.B. Tauris, 1999.

Low, Setha M. and Lawrence-Zuaniga, Denise (eds.). *The Anthropology of Space and Place: Locating Culture*. Boston, MA: Blackwell Publishing Limited, 2003.

Luce, Edward. *In Spite of the Gods: The Strange Rise of Modern India*. New York: Doubleday, 2007.

Macallister, William. "Models of Tribal Culture." Presentation at Center for Advanced Operational Culture Learning. June 20, 2007.

MacLeod, Arlene Elowe. *Accommodating Protest: Working Women, The New Veiling, and Change in Cairo.* New York: Columbia University Press, 1991.

McFarland, Maxie. "Military Cultural Education." *US Army Military Review* (March-April 2005): 62-63.

Maier, Karl. *This House Has Fallen: Nigeria in Crisis.* Boulder, CO: Westview Press, 2003.

Malinowski, B. *Argonauts of the Western Pacific: An Account of Native Enterprise and Adventure in the Archipelagoes of Malanesian New Guinea.* London: Routledge, 1922.

_____. *Magic, Science and Religion.* New York: Doubleday, 1954.

Mamdani, Mahmood. "The Politics of Naming: Genocide, Civil War, Insurgency." *London Review of Books* 29:5, March 2007.

Mandel, Neville. *The Arabs and Zionism before World War I.* Berkeley, CA: University of California Press, 1980.

"Marine Air-Ground Task Force Distributed Operations." *Marine Corps Gazette* Oct 2004, 34-36.

Marine Corps Combat Development Command, "Questions and Answers About Distributed Operations," March 2005: www.mcwl.usmc.mil/SV/DO%20FAQs%2016%20Mar%2005.pdf

Marine Corps Doctrinal Publication 5. Planning. United States Government as Represented by the Secretary of the Navy. July 21, 1997.

Marine Corps Doctrinal Publication 1. Warfighting. United States Government as Represented by the Secretary of the Navy. July 21, 1997.

Marine Corps Warfighting Publication 5-1: Marine Corps Planning Process. Washington, DC: Department of the Navy, 2001.

Mattis, LtGen James N. (USMC) and LtCol Frank G. Hoffman (USMCR, ret). "Future Warfare: The Rise of Hybrid Wars." *US Naval Institute Proceedings*, November 2005.

Mauss, Marcel. *The Gift: Forms and Functions of Exchange in Archaic Societies*. Glencoe: The Free Press, 1954.

McCollam, Douglas. "The Umbrella's Shadow," *Foreign Policy* (online), October 2006

Mclaurin, Melton A. *The Marines of Montord Point: America's First Black Marines*. Chapel Hill, NC: UNC Press, 2007.

Mead, Margaret and Rhoda Metraux. *The Study of Culture at a Distance*. Berghahn Books, [1953] 2000.

Meriwether, Margaret. *The Kin Who Count: Family and Society in Ottoman Aleppo*. Austin, TX: University of Texas Press, 1999.

Michaelson, David. *From Ethnography to Ethnology: A Study of the Conflict of Interpretations of the Southern Kwakiutl Potlatch*. Unpublished PhD Diss. New York: New School for Social Research, 1979.

Monaghan, John and Peter Just. *Social and Cultural Anthropology: A Very Short Introduction*. Oxford, UK: Oxford Univ. Press, 2000.

Monfries, Col Bill. "Initial Officer Training in the Australian Army." In B. Salmoni, comp. and ed. *Pedagogy for the Long War: Teaching Irregular Warfare – Conference Proceedings* (29 Oct – 1 Nov 2007)." Marine Corps Training and Education Command, 2008, 84-85.

_____. "Excerpts from Key-Note Comments," *Pedagogy for the Long War: Teaching Irregular Warfare – Conference Proceedings* (29 Oct – 1 Nov 2007)." Marine Corps Training and Education Command, 2008, 93-97.

Montlake, Simon. "Where the US is Helping to Make Gains Against Terrorism." *Christian Science Monitor*, February 15, 2007:1.

Moore, Jerry D. *Visions of Culture: An Introduction to Anthropological Theories and Theorists*. Lanham, MD: Alta Mira Press, 2004.

Morrison, Kenneth. *Marx, Durkheim, Weber: Formations of Modern Social Thought.* London: Sage Publications, 2006.

Lesch, Ann Mosely. *The Sudan: Contested National Identities.* Bloomington, IN: Indiana University Press, 1999.

Murphy, Liam D. *A History of Anthropological Theory.* Broadview Press, 2003.

Nagl, John. *Learning to Eat Soup with a Knife: Counterinsurgency Lessons from Malaya to Vietnam.* Chicago: University of Chicago Press, 2004.

Nakash, Yitzhak. *Reaching for Power: The Shi'a in the Modern Arab World.* Princeton, NJ: Princeton University Press, 2007.

_____. *The Shi'is of Iraq.* Princeton, NJ: Princeton University Press, 2003.

Nasr, Vali. *The Shia Revival: How Conflicts within Islam Will Shape the Future.* New York: W. W. Norton , 2007.

Norton, Augustus Richard. "Ashura in Nabatiya." *Middle East Insight* 15:3 (May-June 2000): 21-28.

O'Neill, Tom. "Curse of the Black Gold: Hope and Betrayal in the Niger Delta." *National Geographic*, February 2007.

Paret, Peter. *Clausewitz and the State: The Man, His Theories, and His Times.* Princeton, New Jersey: Princeton University Press, 2007.

Patai, Raphael. *The Arab Mind.* Long Island, NY: Hatherleigh Press, [1973] 2002.

Peace Corps Information Collection and Exchange. *Culture Matters: The Peace Corps Cross-Cultural Workbook.* Washington, DC: United States Government Printing Office, n.d.

Perthes, Volker. *The Political Economy of Syria Under Asad.* London: I. B. Tauris, 1997.

_____. "Syria Under Bashar al-Asad: Modernisation and the Limits of Change." *Adelphi Papers* 366 (2004).

Peters, Ralph. "The Human Terrain of Urban Operations," *Parameters.* Spring 2000, 4-12.

Politkovskaya, Anna. *Putin's Russia: Life in a Failing Democracy.* New York: Holt, 2003.

_____. *A Small Corner of Hell: Dispatches from Chechnya.* Chicago: University of Chicago Press, 2003.

Poole, John. *Tactics of the Crescent Moon: Militant Muslim Combat Methods.* Posterity Press, 2004.

"Policing a Whirlwind," *The Economist* 13 December 2007

Pope, Hugh. *Sons of the Conquerors: The Rise of the Turkic World.* New York: Overlook Trade Paperback, 2006.

Proser, Jim, and Jerry Cutter. "I'm Staying with My Boys..." *The Heroic Life of Sgt. John Basilone, USMC.* Lightbearer Communications Company, 2004.

Putnam, Robert. "Diplomacy and Domestic Politics: The Logic of Two-Level Games." *International Organization* 42:3 (1988).

Radcliffe-Brown, A.R. *Structure and Function in Primitive Society.* London: Cohen and West, 1952.

Rashid, Ahmed. *Taliban: Militant Islam, Oil, and fundamentalism in Central Asia. New Have, CT: Yale University Press, 2001.*

Rasmussen, Susan. "Between Several Worlds: Images of Youth and Age in Tuareg Popular Performances." *Anthropological Quarterly* 73:3 (July 2000).

Ravitztky, Aviezer. *Messianism, Zionism, and Jewish Religious Radicalism.* Translated by Michael Swirsky and Jonathan Chipman. Chicago: University Of Chicago Press, 1996.

Record, Jeffrey. *The Specter of Munich: Reconsidering the Lessons of Appeasing Hitler.* Dulles, VA: Potomac Books, 2007.

Rohde, David. "Anthropologists; Scholars embedded in Afghanistan," *The International Herald Tribune,* (October 6, 2007): 3.

Ricks, Thomas. *Making the Corps*. New York: Scribners, 1998.

Rittel, Horst and Melvin Webber. "Dilemmas in a General Theory of Planning," in N Cross (ed), *Developments in Design Methodology*. Chichester, UK: Wiley and Son, 1984.

Roces, Mina. "Kinship Politics in Post-War Philippines: The Lopez Family, 1945-1989." *Modern Asian Studies* 34:1 (February 2000): 181-221.

Rosen, Lawrence. *The Culture of Islam: Changing Aspects of Contemporary Muslim Life*. Chicago: University of Chicago Press, 2002.

Sahlins, Marshall. *Islands of History*. Chicago: University of Chicago Press, 1987.

Salmoni, Barak. "The Fallacy of 'Irregular' Warfare," *Royal United Services Institute Journal*, August 2007.

_____. "Advances in Pre-Deployment Culture Training: The U.S. Marine Corps Approach." *U.S. Army Military Review* (November-December 2006): 79-88.

_____. "Beyond Hearts and Minds: Culture Matters." *Proceedings of the Naval Institute Press*, November 2004.

_____."Debrief of 5th CAG Falluja CMOC Director." Falluja, Iraq, August 2005.

_____. "Historical Consciousness for Modern Citizenship: Egyptian Schooling and the Lessons of History During the Constitutional Monarchy." *In Re-Envisioning Egypt, 1919-1952*, by Goldschmidt, Johnson, and Salmoni, eds. Cairo: American University in Cairo Press, 2005: 164-193.

_____. comp. and ed. *Pedagogy for the Long War: Teaching Irregular Warfare – Conference Proceedings* (29 Oct – 1 Nov 2007)." Marine Corps Training and Education Command, 2008.

Salovey P., and D.J. Sluyter, eds., *Emotional Development and Emotional Intelligence: Educational Implications*. New York: Basic Books, 1997.

Salzman, Philip Carl. *Understanding Culture: An Introduction to Anthropological Theory.* Long Grove, IL: Waveland Press, 2001.

Salzmann, Zdenek. *Language, Culture, And Society: An Introduction to Linguistic Anthropology.* Boulder, CO: Westview Press, 2006.

Sapir, Edward. *Culture, Language, and Personality: Selected Essays.* Berkeley: University of California Press, 1986.

Saville-Troike, Muriel. *The Ethnography of Communication: An Introduction.* Boston, MA: Blackwell Publishing Limited, 2002.

Schechter, Jerrold R. *Russian Negotiating Behavior: Continuity and Transition.* Washington, DC: US Institute of Peace Press, 1986.

Selmeski, Brian R. *Military Cross-Cultural Competence: Core Concepts and Individual Development.* Royal Military College of Canada, Centre for Armed Forces & Society. Occasional Paper Series #1: 16 May 2007.

Service, Elman and Cohen, Ronald. *Origins of the State: The Anthropology of Political Evolution.* Institute for the Study of Human Is, 1978.

Shafqat, Saeed. *Civil-Military Relations in Pakistan: From Zulfikar Ali Bhutto to Benazir Bhutto.* Boulder, CO: Westview Press, 1997.

Sills, Michael D. "Christianimism: Catholic Syncretism in the New World." Southern Baptist Theological Seminary. http://www.sbts.edu/pdf/ChristiAnimism.pdf.

Simpson, E.J. *The Classification of Educational Objectives in the Psychomotor Domain.* Washington, DC: Gryphon House, 1972.

Singerman, Diane. "Where Has All the Power Gone? Women in Politics in Popular Quarters of Cairo." *In Reconstructing Gender in the Middle East,* edited by Fatma M. Göcek and Shira Balaghi. New York: Columbia University Press, 1994, 174-200.

Sluglett, Peter and Farouk-Sluglett, Marion. "The Transformation of Land Tenure and Rural Social Structure in Central and Southern Iraq, c. 1870-1958." *International Journal of Middle East Studies* 15:4 (Nov. 1983), 491-505.

Slyomovics, Susan. "Hassiba Ben Bouali, If You Could See Our Algeria: Women and Public Space in Algeria." *Middle East Report* (January-February 1995): 49-54.

Smith, Daniel Jordan. *A Culture of Corruption: Everyday Deception and Popular Discontent in Nigeria*. Princeton, NJ: Princeton University Press, 2006.

Smith, Larry. *The Few and the Proud: Marine Corps Drill Instructors in Their Own Words*. New York: W. W. Norton, 2007.

Smith, M. Estellie. "The Informal Economy." In Stuart Plattner, ed., *Economic Anthropology*. Stanford, CA: Stanford University Press, 1989.

Smith, General Rupert. *The Utility of Force: The Art of War in the Modern World*. New York : Knopf, 2007.

Spencer-Oatey, H. *Culturally Speaking: Managing Rapport through Talk Across Cultures*. London: Continuum, 2000.

Stack, Carol B. *All Our Kin: Strategies For Survival in a Black Community*. New York: Harper and Row, 1974.

Steele, Robert D. *Information Operations: Putting the "I" Back into DIME*. Carlisle Barracks, PA: Strategic Studies Inst. Of the Army War College, 2006.

Stein, Kenneth W. *The Land Question in Palestine, 1917-1939*. Chapel Hill, NC: University of North Carolina Press, 1984.

Suberu, Rotimi T. *Federalism and Ethnic Conflict in Nigeria*. Washington, D.C.: US Institute of Peace Press, 2001.

Tarabay, Jamie. "What's in a Name? In Iraq, Life or Death." (Audio Recording) *All Things Considered*. National Public Radio, 9 May 2006.
http://www.npr.org/templates/story/story.php?storyId=5394090

Tayler, Jeffrey. *Angry Wind: Through Muslim Black Africa by Truck, Bus, Boat, and Camel*. Boston, MA: Houghton Mifflin, 2005.

_____. "Worse Than Iraq?" *The Atlantic Monthly*, April 2006.

Tentative Landing Operations Manual. Navy Department, 1935.

Terriff, Terry. "Of Romans and Dragons: Preparing the US Marine Corps for Future Warfare," *Contemporary Security Policy* 28:1 (2007).

_____. "Warriors and Innovators: Military Change and Organizational Culture in the US Marine Corps," *Defence Studies* 6:2 (2006).

"The Art of Plain Talk," *Time Magazine* 29 Sept 1967.

Thomas, D.C. , and K. Inkson, *Cultural Intelligence: People Skills for Global Business*. San Francisco, CA: Berrett-Koehler, 2004.

_____, G. Stahl, E. Ravlin, et al., "Cultural Intelligence: Domain and Assessment," un-published M.S., 2007.

Turner, Mark. "The Management of Violence in a Conflict Organization: The Case of the Abu Sayyaf." *Public Organization Review: A Global Journal* 3 (2003): 387-401.

Umstead, Maj Robert and LtCol David R. Denhard, USAF. "Viewing the Center of Gravity through the Prism of Effects-Based Operations." *Military Review* Sept-Oct 2006, 90-95.

United States Army John F. Kennedy Special Warfare Center and School. *Special Operations and International Studies: Political-Military Analysis Handbook*. Fort Bragg, NC, 2004.

United States Joint Forces Command. *Joint Operating Environment: Trends and Challenges for the Future Joint Force through 2030*. Norfolk, VA: USJFCOM, December 2007.

Usunier, Jean-Claude. *Marketing Across Cultures*. New York: Prentice Hall, 1996.

US Marine Corps Combat Development Command and US Special Operations Command Center for Knowledge and Futures. "Multi-Service Concept for Irregular Warfare." Version 2.0, August 2006.

US Marine Guidebook of Essential Subjects. Dept. of Defense Legacy Resource Management Program / Apple Pie Pub, 1998.

Varhola, Major Christopher H. (US Army Reserves). "The Multiple Dimensions of Conflict in the Nuba Mountains of Central Sudan." *US Army Military Review* (May-June 2007): 46-55.

Varhola, Major Christopher H. (US Army Reserves) and Lieutenant Colonel Laura R. Varhola (US Army). "Avoiding the Cookie-Cutter Approach to Culture: Lessons Learned from Operations in East Africa." *Military Review* (December 2006): 73-79.

Von Clausewitz, Karl. *On War.* Edited and Translated by Michael Howard and Peter Paret. Princeton, NJ: Princeton University Press, 1976.

Warren, James A. *American Spartans – The U.S. Marines: a Combat History from Iwo Jima to Iraq.* New York: The Free Press, 1995.

Watanabe, Frank. "How To Succeed in the DI: Fifteen Axioms for Intelligence Analysts." *Studies in Intelligence* (1997) http://www.cia.gov/csi/studies/97unclass/axioms.html.

Weaver, Anne. *Pakistan: In the Shadow of Jihad and Afghanistan.* New York: Farrar, Straus & Giroux, 2002.

Wedeen, Lisa. *Ambiguities of Domination: Politics, Rhetoric, and Symbols in Contemporary Syria.* Chicago, Ill.: University Of Chicago Press, 1999.

West, Deborah L. *Combating Terrorism in the Horn of Africa and Yemen.* World Peace Foundation, March 2005.

West, Harry G. "Girls With Guns: Narrating the Experience of War of Frelimo's 'Female Detachment.'" *Anthropological Quarterly* 73:4 (2000):180-194.

White, Jenny. *Money Makes Us Relatives: Women's Labor in Urban Turkey.* Austin: University of Texas Press, 1994.

Whittle, Richard. "Uncle Sam Wants US Muslims to Serve," *Christian Science Monitor,* 27 Dec 06.

Wise, James E., and Scott Baron. *Women at War: Iraq, Afghanistan and Other Conflicts*. Annapolis, MD: Naval Institute Press, 2006.

Whorf, Benjamin Lee. *Language, Thought, and Reality*. Cambridge, MA: MIT Press, 1956.

Williams, Raymond. *Culture and Society 1780-1950*. New York: Columbia University Press, 1983.

Wolch, J. and Dear, M (eds.). *The Power of Geography: How Territory Shapes Social Life*. London: Unwyn Hyman, 1989.

Woulfe, James B. *Into the Crucible: Making Marines for the 21st Century*. Presidio Press, 1999.

Wunderle, William D. *Through the Lens of Cultural Awareness: A Primer for US Armed Forces Deploying to Arab and Middle Eastern Countries*. Ft. Leavenworth: Combat Studies Institute Press, 2006.

Zerubavel, Eviatar. *Hidden Rhythms: Schedules and Calendars in Social Life*. Berkeley, CA: University of California Press, 1985.

Zerubavel, Yael. *Recovered Roots: Collective Memory and the Making of Israeli National Tradition*. Chicago: University Of Chicago Press, 1997.

Zewde, Bahru. *A History of Modern Ethiopia*. London: James Currey, 1991.

Zuhur, Sharifa. *Revealing Reveiling: Islamist Gender Ideology in Contemporary Egypt*. Albany, NY: State University of New York Press, 1992.